M000026508

Mage Merlin's Unsolved Mathematical Mysteries

The MIT Press
Cambridge, Massachusetts
London, England

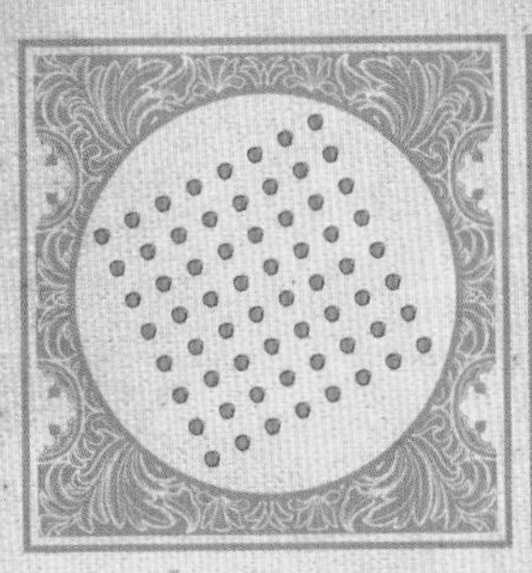

Mage Merlin's Unsolved Mathematical Mysteries

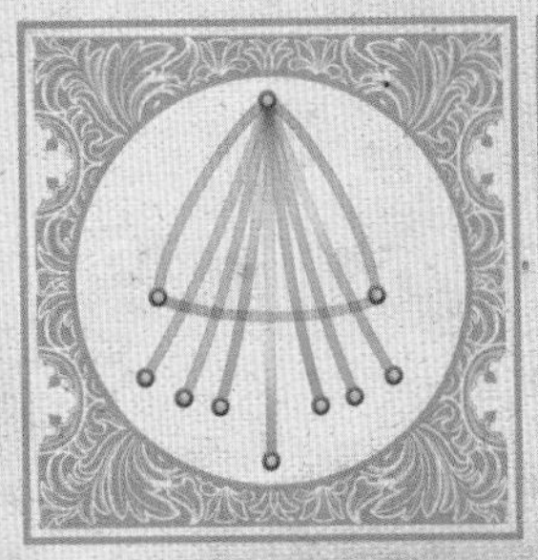

Satyan Linus Devadoss • Matthew Harvey

This book was set in PF Din and Libra by the MIT Press. Printed and bound in the United States of America.

Library of Congress Cataloging-in-Publication Data

Names: Devadoss, Satyan L., 1973– author. | Harvey, Matthew, 1974– author.
Title: Mage Merlin's unsolved mathematical mysteries / Satyan Linus Devadoss and Matt Harvey.
Description: Cambridge, Massachusetts : The MIT Press, [2020]
Identifiers: LCCN 2019044358 | ISBN 9780262044080 (hardcover)
Subjects: LCSH: Mathematical recreations. | Mathematics--Problems, exercises, etc.
Classification: LCC QA95 .D48 2020 | DDC 793.74--dc23
LC record available at https://lccn.loc.gov/2019044358

10 9 8 7 6 5 4 3 2 1

For Jack Morava, a mentor and friend
who instilled in us the joy of playing with big ideas.

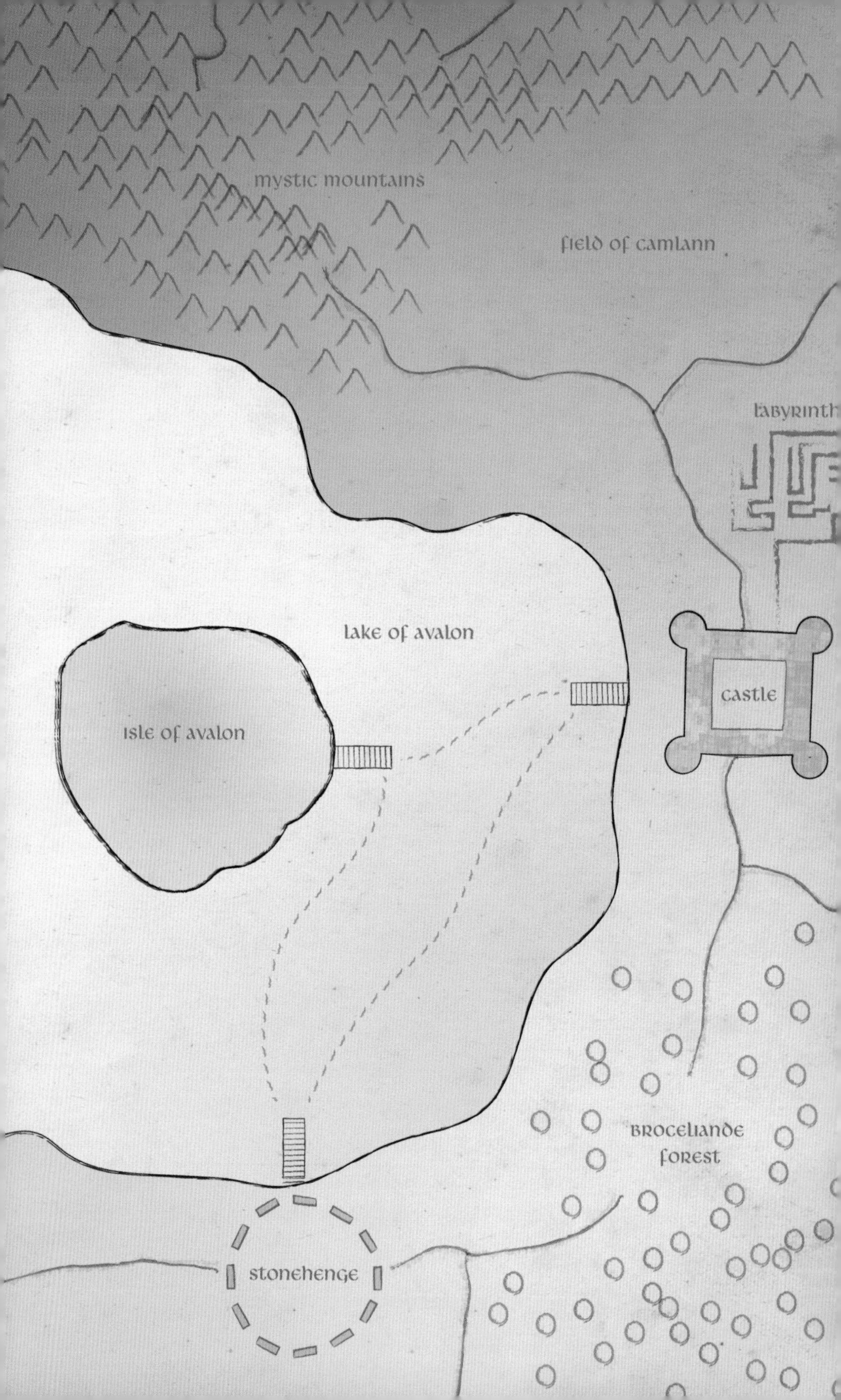
mystic mountains
field of camlann
labyrinth
lake of avalon
castle
isle of avalon
broceliande forest
stonehenge

contents

INTRODUCTION

MARYAM MIRZAKHANI AND THE LANDSCAPE OF MATHEMATICS

Most people think of mathematics as a useful set of tools, designed to answer analytical questions. A typical math education is about learning to use these tools to solve progressively challenging problems. They believe that new ideas in mathematics have been all but exhausted, with its few remaining challenges tucked away in some obscure and technical corner. That is, they view mathematics as a mountain. The wide base is arithmetic, the skills of adding and multiplying that are accessible to everyone. Moving up the mountain, a student discovers algebra, geometry, trigonometry, and eventually calculus and beyond. The journey to the top becomes more and more difficult, with fewer people moving to the next level. Most people can remember exactly where they stopped on this journey, when the climb became too unpleasant due to the altitude. From this viewpoint, the few who make it to the top are the only ones who can appreciate the remaining unsolved challenges of this arcane world.

We view mathematics differently. For us, the joy of mathematics comes not from using a set of tools or manipulating formulas and equations. True delight comes from exploring the unknown frontiers, discovering new connections between ideas, and creating new mathematics of our own. Math is not a mountain but rather an overflowing ice cream cone. The cone is the part of mathematics that is well known and taught in schools. Although the cone tastes fine on its own, its primary purpose is to support the delicious ice cream, the unsolved mathematical mysteries. The higher you move up the cone, the more ice cream there is, so every new level of mathematics opens doors to more unsolved problems. At the very top, vast scoops of unsolved challenges are waiting to be devoured.

Each of the stories in this book is a taste of that ice cream, simple to state yet astoundingly difficult to solve—or, in the spirit of Camelot, a sword in the stone waiting for someone to draw it out, or a spell that no mathematical wizard has yet overcome.

Our fictitious storyteller is Maryam, a distant descendent of Mage Merlin from the Arthurian fables. We chose the name Maryam to honor the real-life genius Maryam Mirzakhani, an Iranian-born mathematician and the only woman to date to win the prestigious Fields Medal, akin to the Nobel Prize of mathematics. Tragically, she died of complications from breast cancer in 2017 at the age of forty. Dr. Mirzakhani had an insatiable curiosity for the beauty and depth of mathematics and was known to spend hours on the floor doodling and drawing diagrams of her ideas. She is the perfect orator for this tale—and a fitting emblem for anyone who delights in the mysteries of mathematics.

For all our modern mathematical wizardry, we may be no closer to solving the problems presented in this book now than Mage Merlin was a millennium ago. Perhaps you will be the one to unlock one of these mysteries, and, like Maryam Mirzakhani, become an extraordinary mage in our own time.

Satyan Devadoss and Matt Harvey

THE MEMORIES OF MAGE MERLIN

My name is Maryam and I'm a mathematician. Let me tell you how that came to be.

When I was young, my grandmother would tell me stories of the magical realm of Camelot. I fell in love with King Arthur, Queen Guinevere, Sir Lancelot, and the rest of the wild and lively characters. But unlike the usual tales of chivalry, jousts, and dragons found in Arthurian legends, my grandma's stories always revolved around fantastic puzzles involving shapes, designs, patterns, and numbers. My favorite character, Merlin the Magician, would make his appearance when things were at their worst, summoned to solve a seemingly impossible problem or unscramble a complicated riddle. Knowing my obsession with Merlin, my grandma would frequently interchange his name with my own. "The Knights of the Round Table can't figure this one out," she would say, with a crinkle in her eyes. "It is time to send for Mage Maryam."

I enjoyed the stories, but it was the puzzles and riddles that really fascinated me. My grandma and I spent hours on end trying to decipher them. I became obsessed with many of them, enlisting the help of any friends or family who were willing to listen. I was able to figure out a few of the simpler puzzles, but most of them eluded me. No matter how frustrated I got or how loudly I complained, my grandma never told me any of the answers. "Sometimes it's healthy to rest in mystery," she would say.

Before she passed away, around the time I was preparing for college, my grandma revealed to me that, according to family lore, we were direct descendants of Merlin himself! Noting my skepticism, she pulled out a tattered leather-bound book and gently placed it in my hands. “This book has been passed down through the generations of Merlin’s lineage,” she told me. “It is the journal of Merlin himself, preserved by his power and magic.” According to her, the Camelot puzzles that so fascinated me when I was younger could all be found within the pages of the journal (although she admitted that the stories she wove around them were mostly products of her own imagination). I was a rational thinker, and it was hard to accept her tale about the book. “Don’t be so eager to dismiss myths, Maryam,” she said. “They hold a hidden layer of truth.”

As I grew older, my curiosity, first fostered by those puzzles, grew. I learned that mathematics gave me a language to play with puzzles, and the freedom to create new ones. In math class, every new idea I learned became a possible tool for me to wield against one of Merlin’s stubborn challenges. That is why I became a mathematician.

* * *

Once I received the leather-bound journal from my grandma, I immediately started to study it in detail. Pushing past the zany characters of Camelot and the far-fetched nature of the stories, I translated the puzzles presented by Merlin into the modern language of mathematics. Some of the riddles were famous problems in mathematics, whereas others were not. Most of them had been solved during the thousand years that passed since they were written down in the journal. However, sixteen of Merlin's puzzles remain—as of now—unsolved. I present them here for your enjoyment.

Before each mystery, I have provided some commentary. The entries themselves, written in Merlin's words, are fairly brief, although I find a certain charm to them. Following each mystery, I have collected the most relevant information I could find regarding the status of the problem: a historical context along with some of the mathematics involved, including current progress toward its solution. Some of the problems were considered by mathematicians quite recently, while others have remained a mystery for hundreds of years.

I don't know how or when these sixteen mysteries might be solved. That is part of what draws me to them: they are questions that don't have easy answers. I have discovered that I am not alone in this. Mathematicians are frequently pulled toward the unknown. For example, in the 1990s, the brilliant mathematician Andrew Wiles solved the most famous problem in mathematics at that time, Fermat's Last Theorem. He had first seen the problem at the age of ten, and its simple statement fascinated him. His childhood love of this puzzle deeply motivated him and fueled his determined pursuit of its solution. I hope that sharing my family stories of Merlin and Camelot will instill this love of mathematics in learners of all ages.

MYSTERY 1

2

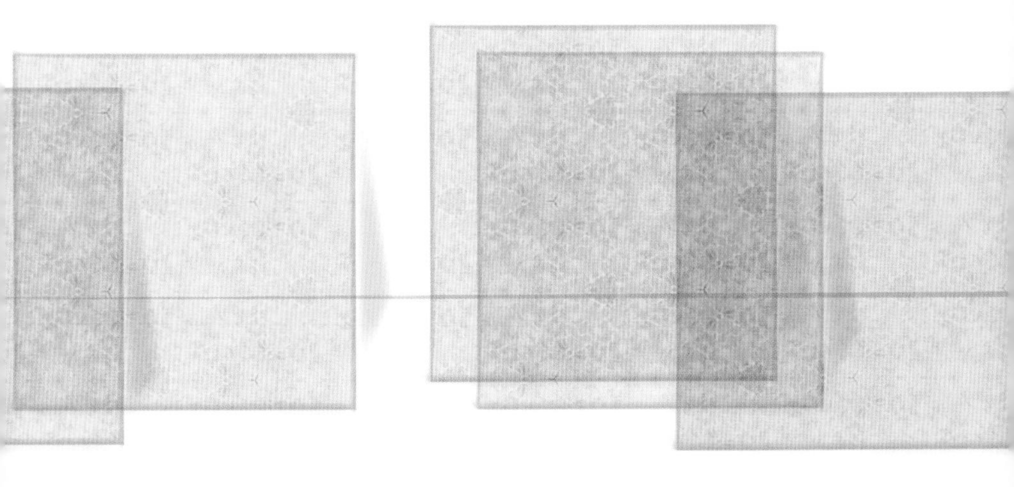

Great Hall Window

This first challenge is one of my favorites, where we find Merlin trying to cover a large square with six smaller ones. Throughout my education, I have seen what an effective tool mathematics is for solving problems—helping a rover to land on Mars, a computer to outthink chess masters, a phone to navigate us around the world. So it amazes me that for all its successes, mathematics still can't answer such a simple question.

I was summoned to Camelot in the middle of the night.

The full moon shone brightly upon the majestic castle. Arriving in the Great Hall, the resting place of the mighty sword Excalibur, I noticed that the resplendent stained glass window, a perfect 201 × 201 square, lay shattered on the floor.

There were six 100 × 100 square tiles nearby, each inlaid with an ancient design. King Arthur asked if I could cover the window opening using these tiles, protecting Excalibur from the dangers outside. I could arrange and overlap the six sacred tiles, but certainly not break them.

All night I toyed with the tiles, positioning and repositioning them in various configurations, but even with my powers of magic and logic, I could not succeed.

Four 1×1 squares fit together to exactly cover a 2×2 window. If this square window is slightly larger, the four squares will not be enough to cover it. In 2000, Trevor Green showed how to use seven tiles to cover a slightly larger square.

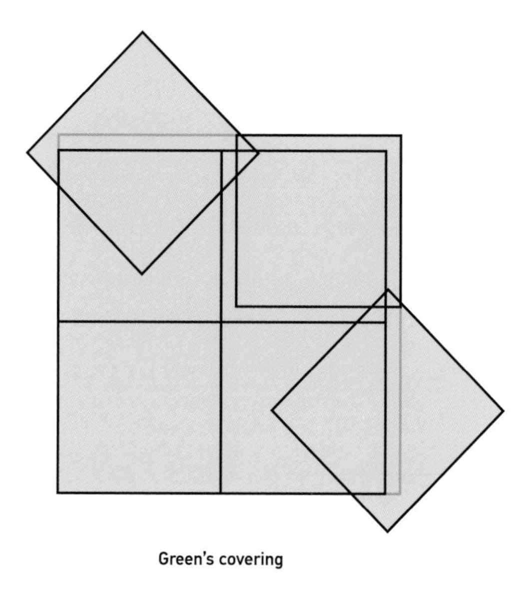

Green's covering

In 2009, Janusz Januszewski proved that five squares are not enough to cover the square window, no matter how they are arranged. As Merlin suggests, the question of whether six squares is enough remains unsolved. The six 100×100 tiles have a total area of 60,000, significantly more than the area of the 201×201 window, which is 40,401. However, it seems impossible to position the first few tiles so that they do not leave long narrow gaps.

Although Merlin's problem is quite specific, it is one of a number of similar questions: What is the minimum number of 1×1 squares that are needed to cover an $n \times n$ square? How many squares are needed to cover rectangles or triangles of various sizes? Given a fixed number of tiles, what is the largest square that they can cover? In 2005, Alexander Soifer conjectured the following:

UNSOLVED: The largest square that can be covered with $n^2 + 1$ squares of size 1×1 will be of size $n \times n$.

Definitive answers to these questions are few and far between. Computers have had some success finding efficient covers, but even in those cases, it is not known whether the solutions are optimal.

MYSTERY 2

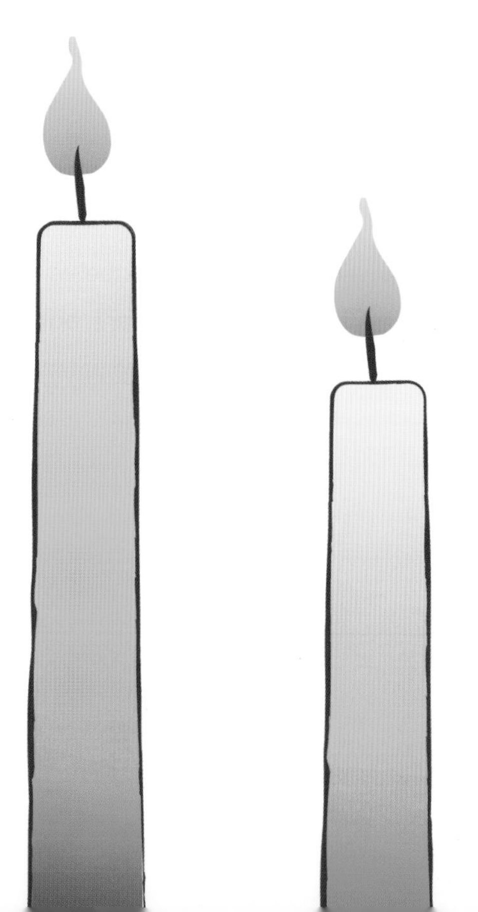

Blessed Birthday Banquet

One of my earliest mathematical fascinations was with prime numbers. A number is *prime* if it is evenly divisible only by 1 and itself. Thus, numbers like 3, 7, and 11 are prime, but 4, 9, and 15 are not. As the fundamental building blocks of multiplication, prime numbers have fascinated mathematicians for thousands of years. Finding patterns among primes has led us to develop entirely new techniques and even new branches of mathematics, and yet many questions remain unanswered.

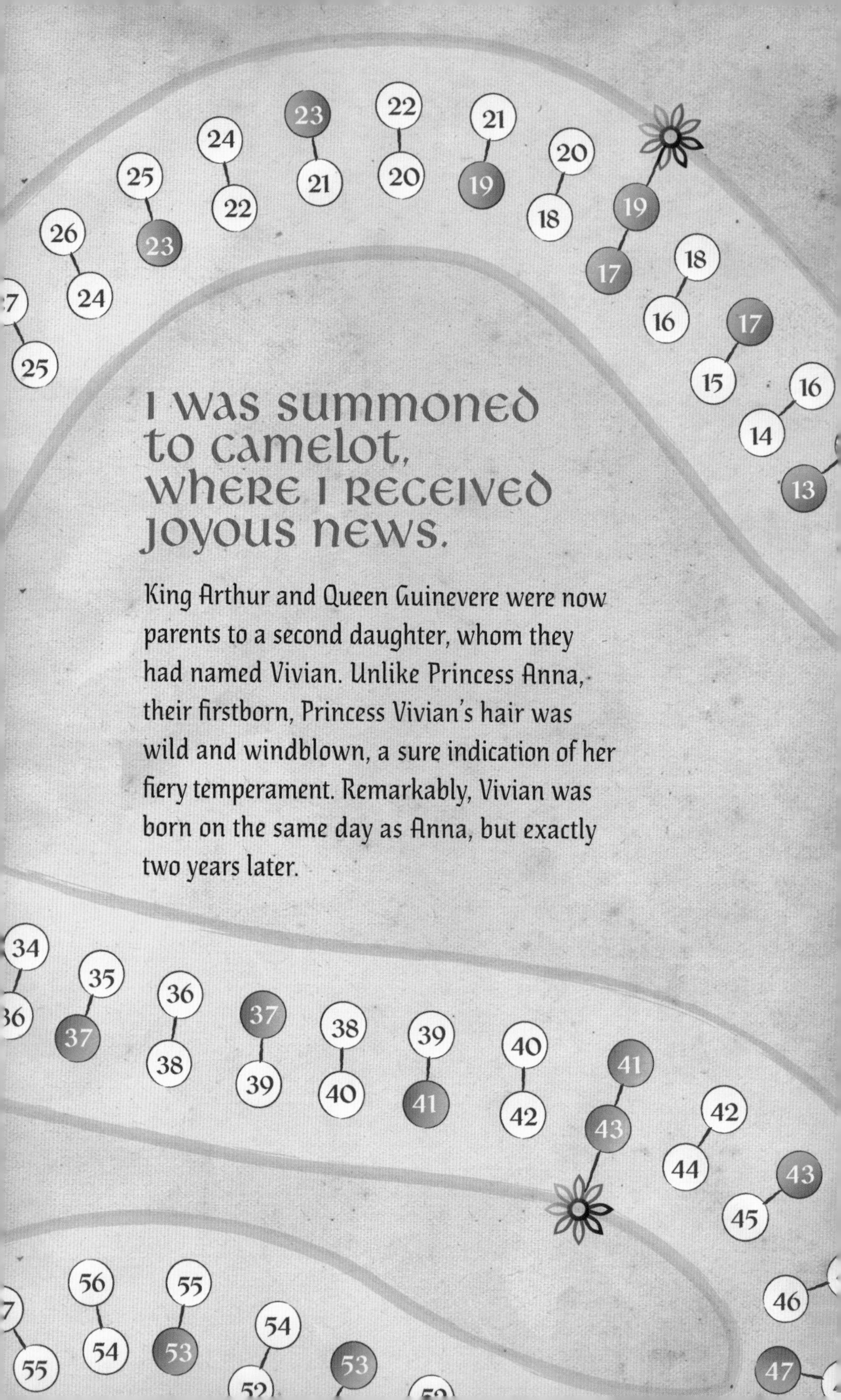

I was summoned to Camelot, where I received joyous news.

King Arthur and Queen Guinevere were now parents to a second daughter, whom they had named Vivian. Unlike Princess Anna, their firstborn, Princess Vivian's hair was wild and windblown, a sure indication of her fiery temperament. Remarkably, Vivian was born on the same day as Anna, but exactly two years later.

To celebrate this marvelous blessing, the king and queen decided to enact a new tradition in Camelot: every year on their shared birthday, each daughter would light a red candle if turning a prime age and a white candle otherwise. On the years when both daughters lit a red candle, a grand banquet would be held throughout Camelot.

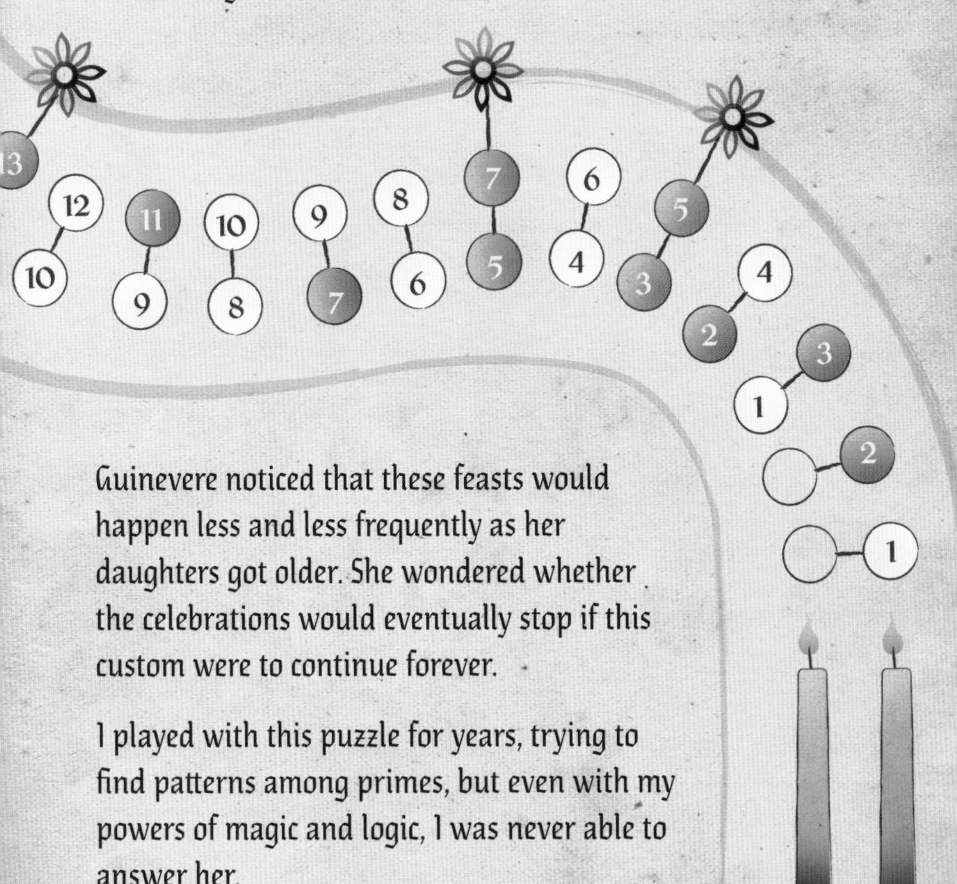

Guinevere noticed that these feasts would happen less and less frequently as her daughters got older. She wondered whether the celebrations would eventually stop if this custom were to continue forever.

I played with this puzzle for years, trying to find patterns among primes, but even with my powers of magic and logic, I was never able to answer her.

One of the earliest findings on primes comes from the great Greek geometer Euclid. Around 300 BCE, Euclid compiled much of what was known about mathematics at the time into a book called *The Elements*. In it, he provided the first known proof that there are infinitely many prime numbers. Euclid argued that if there were only finitely many primes, then it would be possible to multiply all the primes together and then add one. He was able to show that the resulting number would be neither prime nor not prime. Since that is impossible, Euclid concluded that the set of primes could not be finite.

The story of the princesses' birthdays involves pairs of primes that differ by two. These are called *twin primes*. They occur frequently among small numbers: 3–5, 5–7, 11–13, 17–19, but less frequently among larger numbers. Guinevere's question is a natural follow-up to Euclid's result: since there are infinitely many primes, are there infinitely many twin primes? The question was formally posed by Alphonse de Polignac in 1846 as a conjecture:

UNSOLVED: There are infinitely many twin primes.

For a long time, little progress was made toward proving or disproving the conjecture. But in 2013, a previously unheralded mathematician named Yitang Zhang surprised the mathematical world with a major breakthrough. Zhang established the first finite bound on gaps between primes, proving that there are infinitely many prime pairs of the form $(p, p + N)$, where N is some number less than 70 million.

Following a flurry of activity, the upper bound N was significantly narrowed to be less than 246, yet it remains to be seen whether it reaches 2, thereby solving the twin prime conjecture.

MYSTERY 3

Tinkering Toy Trouble

This story of toys and colors involves the mathematics of graphs. A *graph* captures the relationships between pairs of objects, visualized as points connected by lines. They are used in a vast range of areas, from linguistics and chemistry to computer science and network analysis. At first glance, graphs seem to be simple objects. Merlin teases out their complexity in unexpected ways by entering the world of children's tinker toys.

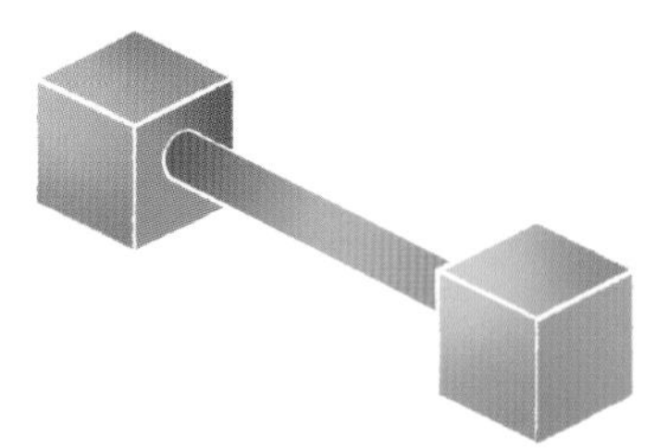

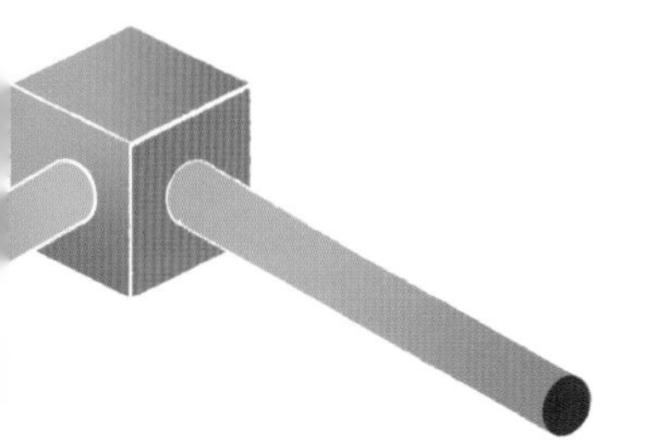

I was summoned to Camelot for a most childish affair.

It was exactly eleven years after the birth of Princess Vivian, and a grand banquet celebration had just ended. For their birthday, the two princesses received an abundance of gifts. But the most interesting was a set of toys from the sorceress Morgana, whose power and brilliance nearly matches my own.

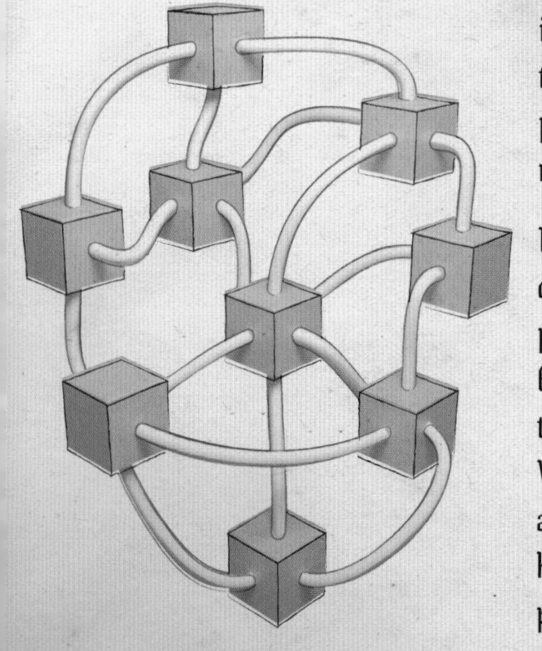

Morgana's wonderfully crafted construction set was made of perfect cubes and flexible tubes. Each tube was used to connect the sides of two distinct cubes. Vivian's toy set was unpainted and could be enjoyed freely, but her older sister Anna's set was painted in eight distinct colors, and required an additional rule of play: for any three pieces appearing in a row, each must be a different color.

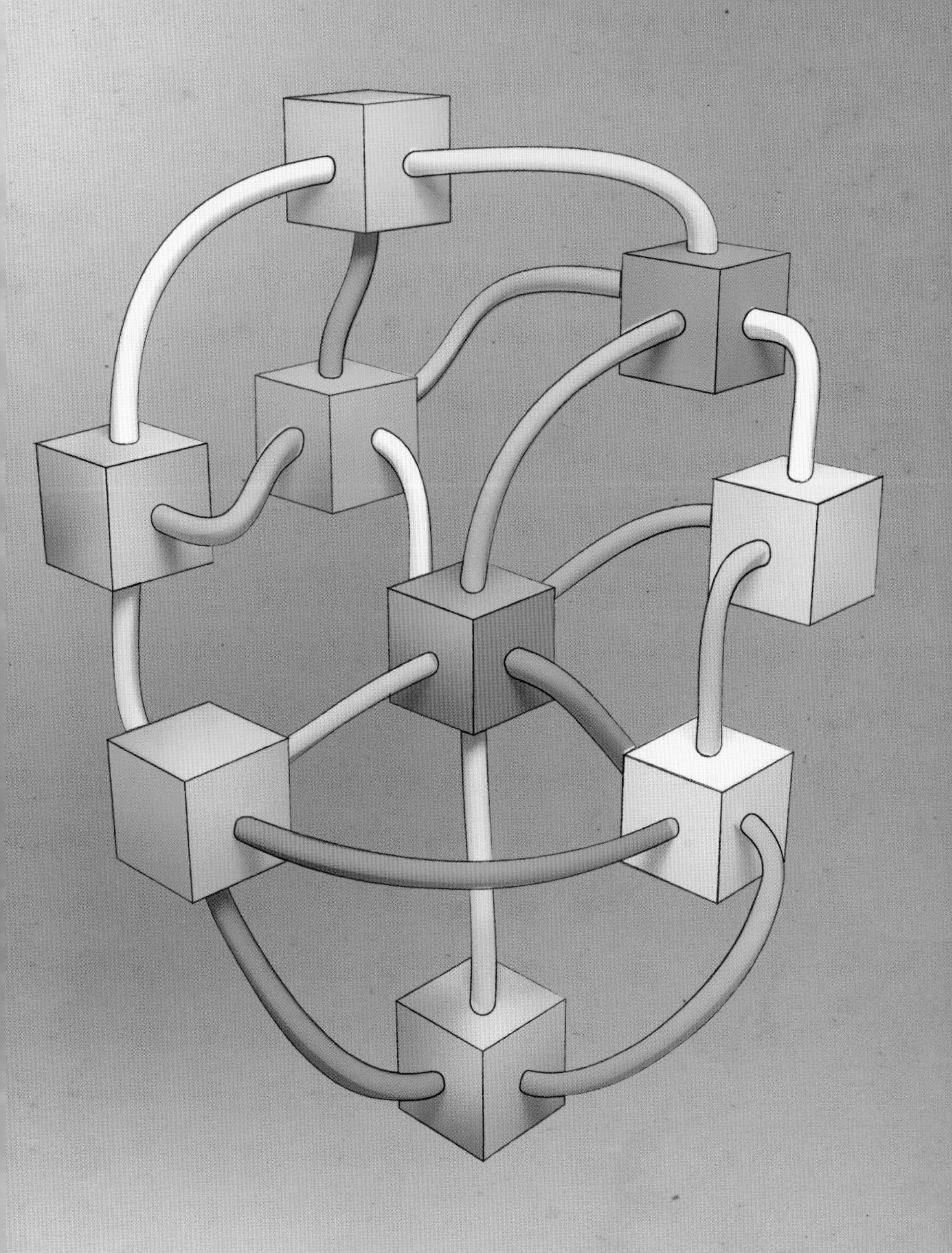

To challenge my wit, Morgana asked whether any model Vivian built with the unpainted set could be recreated using Anna's colored set and its rule.

I tinkered with these toys, trying to make this mystery tractable, but even with my powers of magic and logic, I was never able to answer her.

A construction built from Vivian's unpainted set is a physical representation of a graph. For this story, the cubes in the construction set are the points (or nodes) and the tubes are the lines (or edges) that connect pairs of nodes. Since cubes have six sides, the graphs that Vivian can create will never have more than six edges connected to any one node.

Anna's challenge is to find a way to color the nodes and edges so that three consecutive objects, whether node-edge-node or edge-node-edge, are of three different colors. Anna's kit gives her eight different colors to work with.

When coloring any graph with these rules, one thing becomes clear: if a node has E edges emerging from it, then at least $E + 1$ colors will be needed to successfully color the node and its attached edges: one color for the node, and E more colors for each edge, in order to satisfy the edge-node-edge color rule. However, color choices made around this one node impact the possible choices at neighboring nodes. And choices made at those nodes affects their neighbors, creating rippling constraints that reverberate through the rest of the graph.

If E is the maximum number of edges connected to any single node of a graph, it is certainly reasonable to guess that some graphs would require substantially more than $E+1$ colors. It is known that every graph can be colored with at most $2E+2$ colors. But in the 1960s, Mehdi Behzad conjectured a much stronger result:

UNSOLVED: If E is the maximum number of edges connected to any single node of the graph, and T is the minimum number of colors required to color it, then $T \leq E + 2$.

Currently, the conjecture is known to be true for all graphs where $E \leq 5$. Morgana's challenge deals with the next case: when $E = 6$, are $T = 8$ colors enough?

MYSTERY 4

Glorious Gift Wrappings

Some of the most beautiful geometric objects we encounter in our lives are polyhedra. A *polyhedron* is a 3-D solid with flat sides made of polygons. A cube is one of the more famous examples. In this puzzle, Merlin is worried about skillfully using gold paper to wrap polyhedra of all shapes and sizes. It's a beautiful example of the interplay between 3-D objects and the 2-D paper used to cover them.

I was summoned to Camelot, for the vernal equinox was fast approaching.

Camelot became flooded with hundreds of celebratory gifts during this festive season. Traditionally, each present was fully enclosed in a container made of flat wooden panels. The resulting packages were of all shapes and sizes, with the wooden panels meeting in all kinds of interesting angles.

For a gift to be presented to the king and queen, however, the wooden package had to be completely covered in pure gold foil. This year, the tedious gift-wrapping task fell upon Sir Dürer.

For each package, Dürer wanted to cut out one piece of golden foil: this would then be folded over the package *perfectly*, fully covering the wooden panels but without the foil overlapping itself. I was asked if this would always be possible, no matter the shape of the package.

I tried to wrap my mind around this during the weeks leading up to the equinox, but even with my powers of magic and logic, I was not able to answer him.

the wrappings of three packages

Merlin wants to know whether he can fold a flat piece of paper perfectly around a polyhedron: covering all of its faces without overlapping any of the paper. Seen in reverse, his question asks whether it is possible to cut along the faces of a polyhedron so that it unfolds into a single flat piece that does not overlap itself.

UNSOLVED: Every polyhedron can be cut and unfolded into a single flat surface without overlap.

This is always possible for a convex polyhedron. A polyhedron is *convex* if there exists an unbroken line-of-sight between any two points inside it. For example, a cube is convex but a cube with a smaller cube poking out of the middle of one face is not. It is the indentations around the smaller cube that prevent it from being convex.

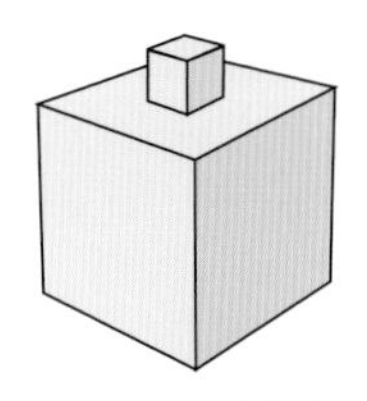

a nonconvex polyhedron

Here's how to always unfold a convex polyhedron: make cuts from a point *P* on the surface of the polyhedron to each of its corners, taking the shortest path along the surface from *P* to each corner. Then peel open the polyhedron along the cuts to lay it down flat. If the polyhedron is convex, the flaps of the unfolded shape will not overlap. This method even works for some nonconvex polyhedra, but unfortunately not all.

There is a related question inspired by the sixteenth-century artist Albrecht Dürer: which polyhedra can be unfolded if we permit cuts only along the edges? A polyhedral *net* is obtained by cutting some edges of a polyhedron so that it unfolds into a single flat piece that does not overlap itself. In his remarkable book *Underweysung der Messung mit dem Zirckel und Richtscheyt* (1525), Dürer presented nets for several polyhedra, one example being that of the truncated tetrahedron.

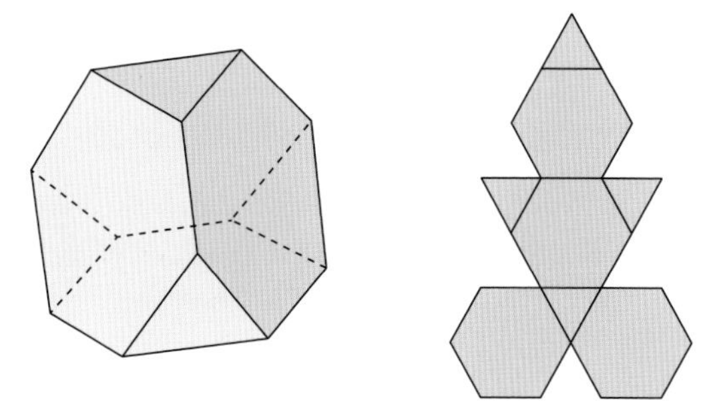

a truncated tetrahedron (left) and its net (right)

Unfolding polyhedra along edges works not just for the truncated tetrahedron, but for numerous other polyhedra. In 1975, Geoffrey Shephard wondered if this was always possible for convex polyhedra. Every convex polyhedron mathematicians and computer scientists have been able to imagine so far have produced nets, but whether this is always possible remains tantalizingly open.

UNSOLVED: Every convex polyhedron can be cut along some of its edges to form a net.

MYSTERY 5

perfect cake

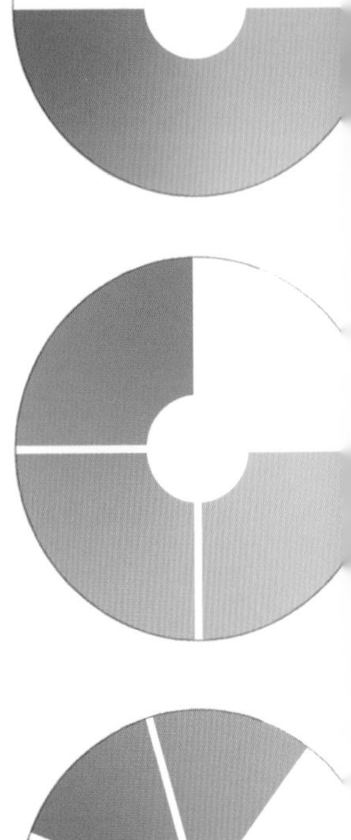

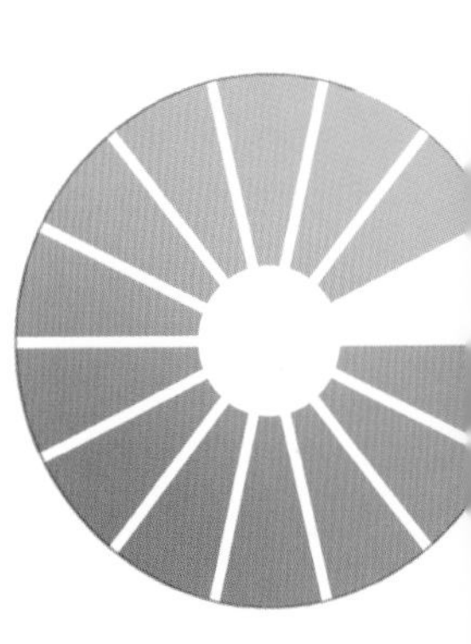

COMBINATIONS

Did the Camelot royalty ever tire of their strangely impractical celebrations? Merlin's next adventure deals with cakes that are cut not according to the number of guests who attend, but based on the *factors* of the anniversary year—all the numbers that divide evenly into it. For example, the factors of 6 are 1, 2, 3, and 6. Some of the most venerable questions in mathematics have to do with factorization, the study of how numbers can be written as products of other numbers.

I was summoned to Camelot on an anniversary of its founding.

A festival had been established to commemorate the day Arthur drew the sword Excalibur from the stone and began his reign as king. For the jubilee, Camelot's bakers prepared many varieties of delicious and beautifully decorated cakes for the king and his guests, each cake representing one factor of the anniversary year. The cakes were then cut into as many slices as the factors they represented.

Since 1 is a factor of every number, there was always a whole uncut cake, which was given to me, as the chief architect and steward of Camelot. King Arthur, however, received exactly one slice from each of the cut cakes.

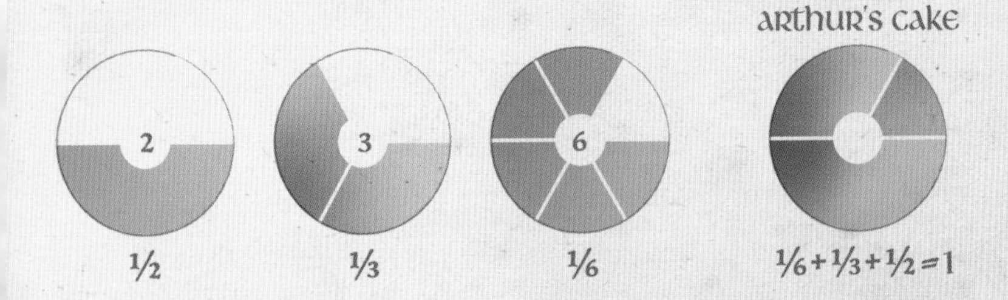

Sometimes Arthur's slices added up to a perfectly whole cake (sixth anniversary), but most of the time his slices added up to more than an entire cake (twelfth anniversary) or less than an entire cake (sixteenth and twenty-first anniversaries). Arthur noticed that every time he received a perfectly whole cake, half of it came from one single slice.

He asked whether it was possible for him to receive a perfectly whole cake without having a slice that is half the size.

I chewed on his question for years, but even with my powers of magic and logic, I was never able to answer him.

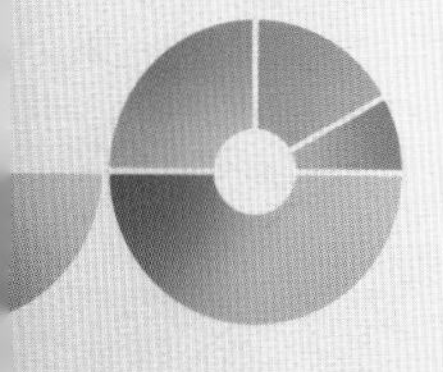

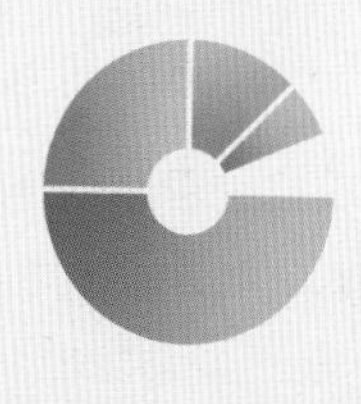

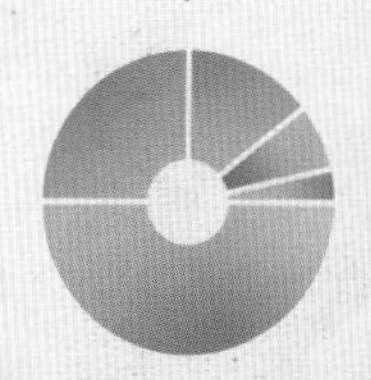

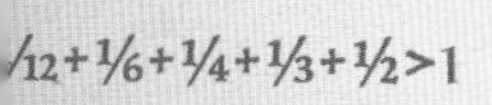

$\frac{1}{16}+\frac{1}{8}+\frac{1}{4}+\frac{1}{2}<1$

$\frac{1}{21}+\frac{1}{7}+\frac{1}{3}<1$

$\frac{1}{28}+\frac{1}{14}+\frac{1}{7}+\frac{1}{4}+\frac{1}{2}=1$

Merlin's quest for the perfect odd cake describes a mathematical question involving what are called perfect numbers. A number is *perfect* if it is equal to the sum of its proper factors. As an example, the *proper* factors of 6 are 1, 2, and 3. (A number is always a factor of itself, but is not considered a proper factor.) Since $6 = 1 + 2 + 3$, 6 is a perfect number. It is because 6 is perfect that Arthur's cake is perfectly whole: adding up the sizes of the pieces gives

$$\frac{1}{2}+\frac{1}{3}+\frac{1}{6}=\frac{3}{6}+\frac{2}{6}+\frac{1}{6}=\frac{(3+2+1)}{6}=\frac{6}{6}=1.$$

Perfect numbers are actually quite rare. The only perfect numbers less than 100 are 6 and 28. More commonly, a number is deficient, meaning the sum of its proper factors is less, as with 8:

$$1+2+4<8;$$

or abundant, meaning the sum of its proper factors is more, as with 12:

$$1+2+3+4+6>12.$$

A number is more likely to be deficient than abundant, but there are both infinitely many deficient numbers and infinitely many abundant numbers. But the following still remains a mystery:

UNSOLVED: There are infinitely many perfect numbers.

Merlin's question is whether a perfect number can ever be odd, pointing to the following conjecture:

UNSOLVED: All perfect numbers are even.

Computers can now check whether fairly large odd numbers are perfect, and to date, none have been found. In contrast, even perfect numbers, while rare, are better understood. Euclid's *The Elements*, written more than two millennia ago, documents an important connection between even perfect numbers and prime numbers: if a number of the form $q = 2^n - 1$ is prime, called a Mersenne prime, then $q\,(q + 1)/2$ is a perfect number. For instance, when $n=2$, $q = 2^2 - 1 = 3$ is the first Mersenne prime and $q\,(q + 1)/2 = (3 \times 4)/2 = 6$ is the first perfect number. When $n = 3$, $q = 2^3 - 1 = 7$ is the second Mersenne prime and $q\,(q + 1)/2 = (7 \times 8)/2 = 28$ is the second perfect number.

More recently, Leonhard Euler proved in the nineteenth century that *every* even perfect number has this form. Many numbers of the form $2^n - 1$ are not prime and therefore do not lead to perfect numbers; since the advent of computers, however, the search for ever larger Mersenne primes has reached new heights. As of 2018, the largest known Mersenne prime has over 24 million digits. However, Mersenne primes will only ever lead to even perfect numbers, and so any odd perfect, if it exists, will have to be found by some other method.

MYSTERY 6

Round Table Tiles

Merlin has previously pondered covering objects with shapes, and now he turns to tiling with shapes. All of us encounter tilings frequently, from simple bathroom floors and kitchen backsplashes to intricate friezes and tapestry inlays. Whereas all these are inevitably finite in size, mathematicians envision tilings that stretch on forever, covering an infinite plane. Despite the difference in scale, real-world tilings do give a proper sense of the pattern and symmetry of their mathematical brethren. In this story, Arthur asks Merlin to cover a table with tiles in a way that lacks the usual symmetry.

I was summoned to Camelot for decorating advice.

The Round Table was a renowned symbol of unity and equality in Camelot. Although immense in size, it was a simple wooden table, so King Arthur wanted to adorn it by covering the table with an intricate tiling pattern.

Simple tiles were first considered, such as two-colored squares. These squares could be used to create a *symmetrical* pattern, which would repeat moving from one square to another. Or the squares could be used to create a *nonsymmetrical* pattern. For instance, one misplaced square tile would shatter the symmetry, making the table look different depending on where one sat.

Recently, Arthur had learned of Sir Penrose, a knight from a nearby kingdom who had created a tiling masterpiece. The beauty of Penrose's work was not in the tiling but his two special tiles: whereas the two-colored square tiles could be arranged to yield either symmetrical or nonsymmetrical patterns, Penrose showed that every possible tiling made from his two special tiles will always be nonsymmetrical.

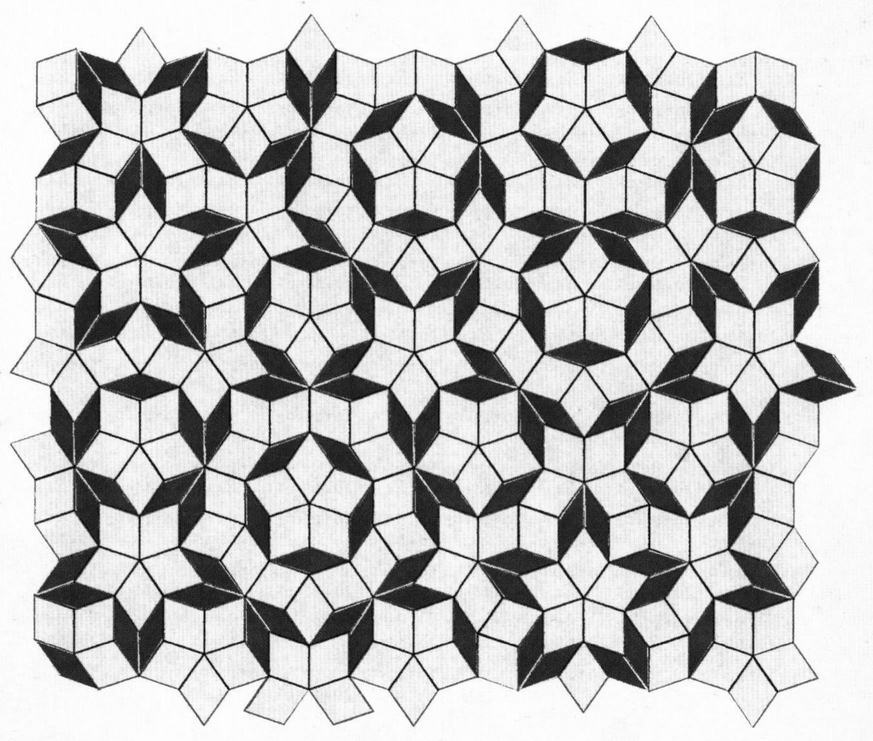

To symbolize the unity of Camelot, Arthur wanted one special tile that did what the two tiles of Penrose did: a single tile that would always produce nonsymmetrical patterns regardless of how it was placed on the Round Table.

I delved into designing such a dazzling tile, but even with my powers of magic and logic, I was never able to succeed.

Most of the tilings that we encounter regularly, in either the real or the mathematical realm, are *periodic*—in other words, they have translational symmetry: if a copy of the tiling is shifted by just the right amount, it will line up exactly with the original. The symmetry is appealing, both from a manufacturing perspective, as in bathroom tiles, and from an aesthetic one, as in friezes. But not all tilings are periodic.

In the story, Merlin describes a tiling whose symmetry is frustratingly broken because a square tile has been set out of place. In that case, it is apparent that the problem is the shoddy handiwork of the tile setter rather than a defect of the tile itself. This is generally the case. Many erratic tilings just need a bit of rearrangement to become periodic ones. And yet, one wonders, are there tiles that by their very shape cannot be placed periodically? Such tilings are *aperiodic*.

The challenge to finding an aperiodic tiling is twofold. First, the shapes must tile the plane, which is already difficult. Second, and even more difficult, something fundamental about the shapes of the tiles must prevent them from being placed periodically. It is perhaps a bit surprising, therefore, that aperiodic tilings have actually been found.

The first aperiodic tiling, which used 20,426 distinct tile shapes, was discovered in 1964 by Robert Berger. Since then, aperiodic tilings requiring fewer tile shapes have been found. In 1973, the physicist Roger Penrose reduced this to two shapes. Using these shapes, he actually found two different intertwined aperiodic tilings, the kite-and-dart and the rhomb tilings (the latter of which appears in Merlin's story). Typically, Penrose's tiles have markings on them that specify how the tiles can be placed with respect to one another. Alternately, the edges of the tiles can be notched to enforce the same ruleset.

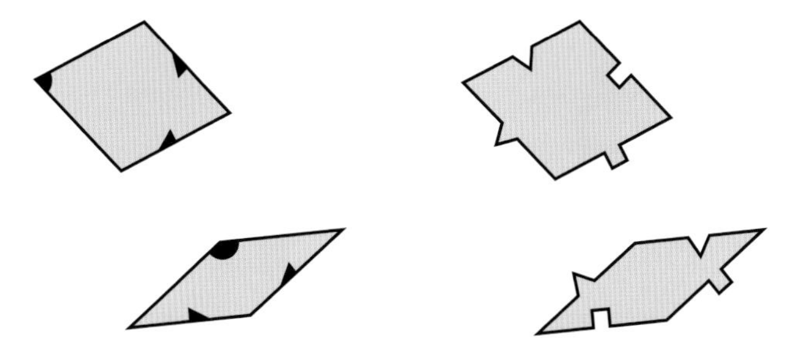

Penrose's marked (left) and unmarked (right) tiles. Marked corners can only meet other marked corners. Marked edges can only meet other marked edges, and the two marks must align to form a triangle (rather than a parallelogram).

Penrose's proof that his shapes can tile a plane but cannot tile it periodically is a subtle argument that depends on the intricate connection between the kite-and-dart tile and the rhomb tile. Since Penrose's work, mathematicians have sought a *single* connected shape that will tile a plane aperiodically. None has been found.

UNSOLVED: There is a single connected tile that tiles a plane aperiodically.

MYSTERY 7

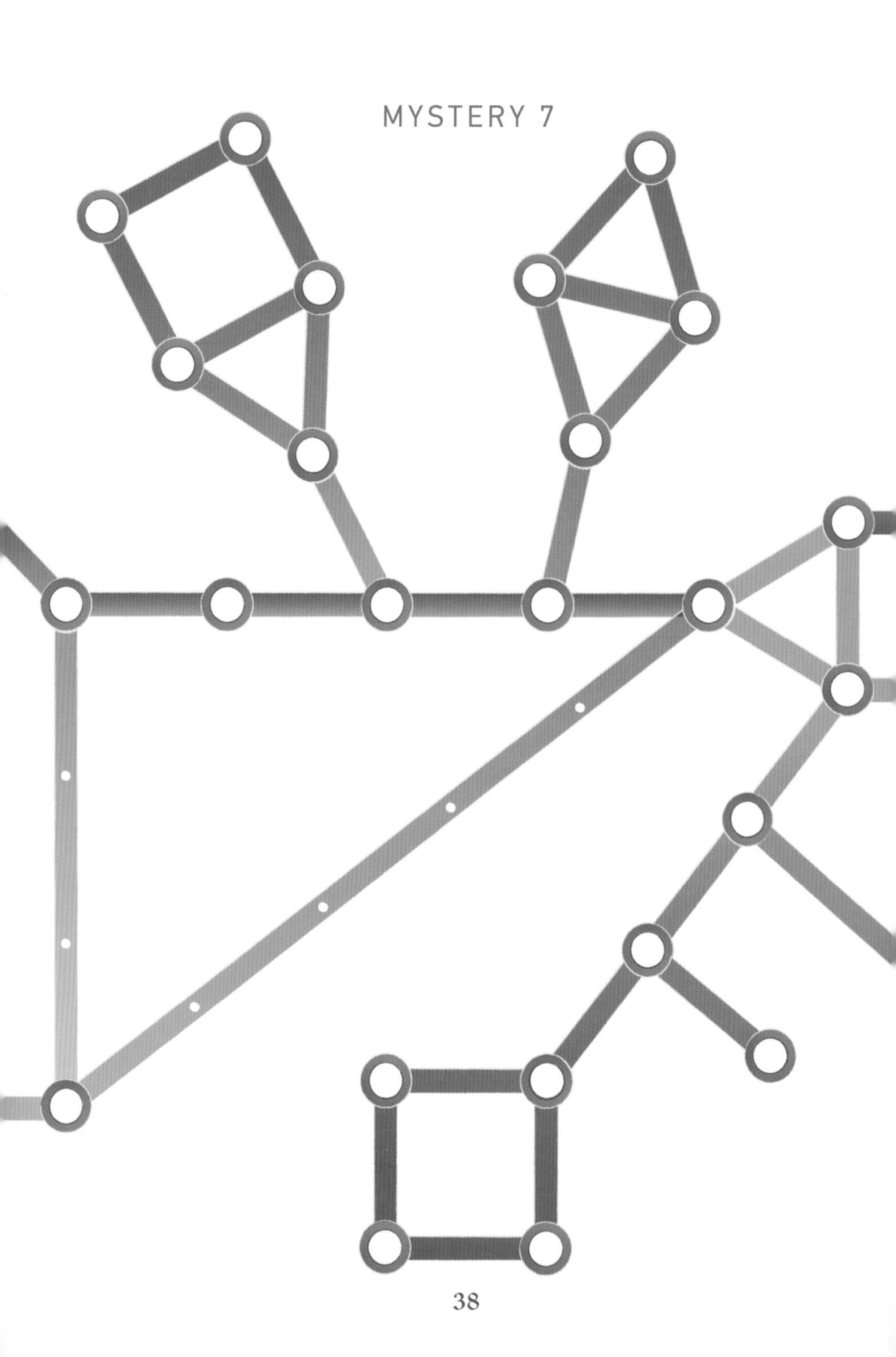

Lancelot's Labyrinth Labors

This mystery combines my two favorite things: mathematics and doodling. If a graph can be drawn on paper so that none of its edges cross, it is called *planar*. There are many ways to draw planar graphs, to place the dots on paper and then to connect them with edges that don't cross. This adventure finds Merlin curious about whether he can draw planar graphs with perfectly straight lines whose lengths are all whole numbers. Along the way, he makes note of Guinevere's interest in Lancelot.

I was summoned to Camelot to design a devious maze.

Lancelot had shown bravery in battle with sword and spear, but for him to become a legendary Knight of the Round Table, one final challenge awaited him: a labyrinth. Traditionally, a labyrinth consisted of straight, flat paths that met only at circular junctions. Plans were being drafted for a cunning labyrinth, one where each path presented a deadly obstacle for Lancelot to overcome.

Queen Guinevere asked me whether the final labyrinth blueprint could be altered so that the length of each path would be an exact multiple of ten, her favorite number. She wanted the same set of straight paths to meet at each junction, as drawn in the blueprint, but allowed me to alter the *positions* of the paths and the junctions.

I laid aside my curiosity regarding Queen Guinevere's unusual interest in Lancelot, and tried to find my way through this maze. But even with my powers of magic and logic, I was not able to answer her.

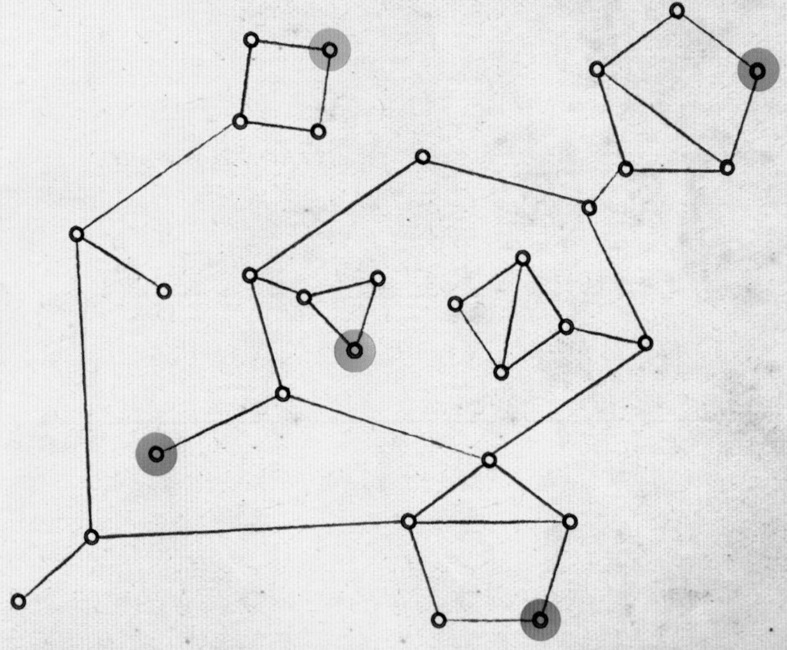

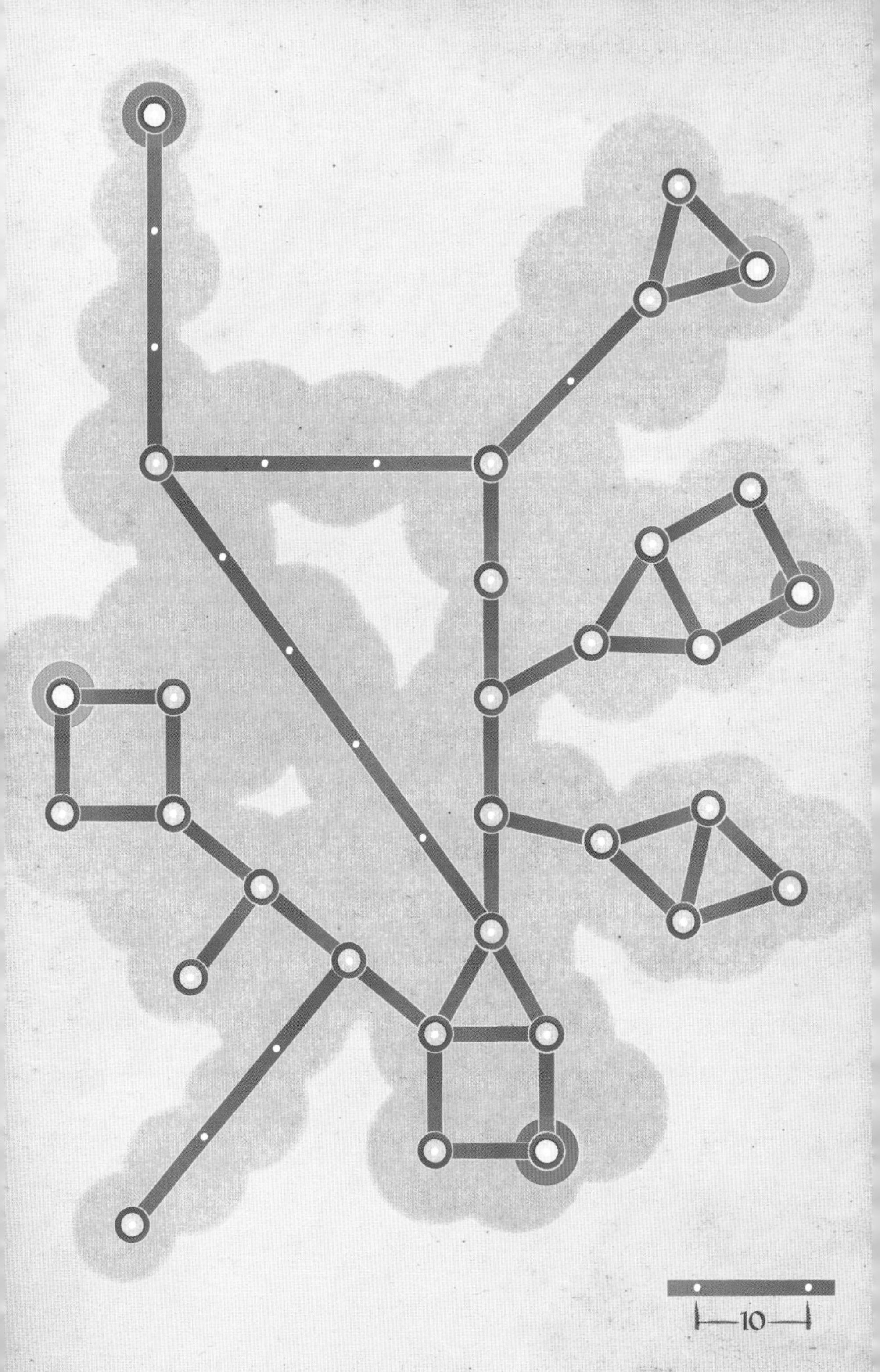
10

This is another story using graph theory, now focusing on planar graphs. For example, the five-pointed star graph is planar because it can be redrawn as a pentagon. The K_5 graph, consisting of five nodes and an edge connecting every pair of nodes, is not planar.

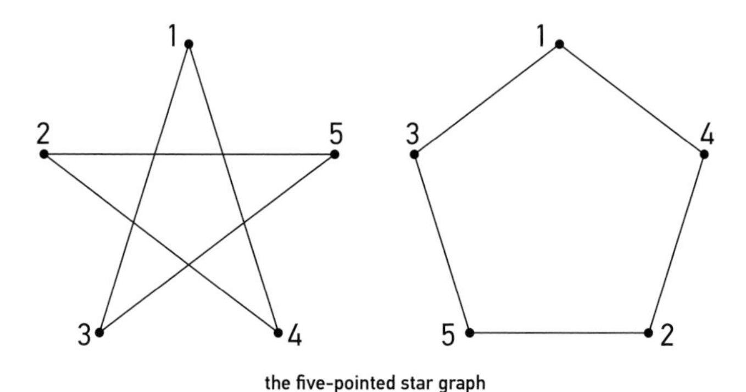

the five-pointed star graph

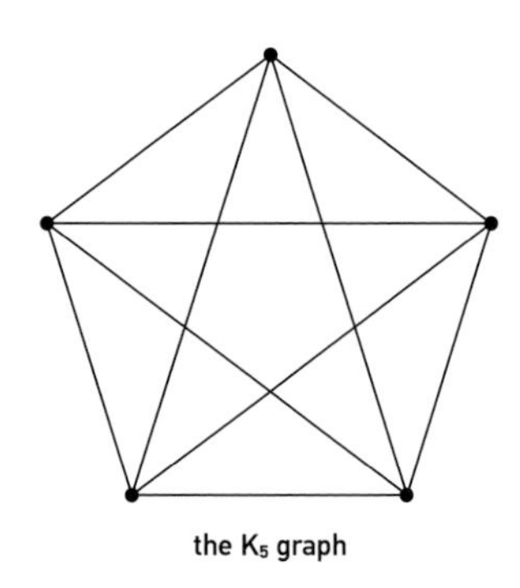

the K_5 graph

The edges of a planar graph are often drawn with curved edges for aesthetic or practical reasons, but in the mid-twentieth century, Istvan Fary, Klaus Wagner, and Sherman Stein independently proved that this is not truly necessary. If a graph is planar, it can be drawn using only straight line segments, a result that is now called Fary's theorem. In the 1980s, Heiko Harboth conjectured something more:

UNSOLVED: Every planar graph can be drawn with segments whose lengths are whole numbers.

In our story, the knight's plan for the labyrinth describes a planar graph, with circular junctions as nodes and dangerous paths as edges. If Merlin is sure that he can redraw the graph with whole number side lengths, he can then scale it up by a factor of ten to create a blueprint for the desired maze.

Harboth's conjecture has not been proved in this general context, but it has been confirmed for some special classes of graphs, including bipartite graphs, whose nodes can be divided into two sets so that every edge connects a node of one set to a node of the other, and 3-regular (or cubic) graphs, which have exactly three edges meeting at each node.

MYSTERY 8

stonehenge tapestries

Merlin is back to graphs again, a recurring topic in his journal. But unlike the last mystery, where Merlin was asked to think about all kinds of planar graphs, this exploit feels more reasonable: find one graph with a special property. This was always one of my favorite puzzles. I filled pages and pages in my notebooks with scribbles of graphs, trying to find that special one. Notably, this entry also mentions the mysterious prehistoric monument of Stonehenge.

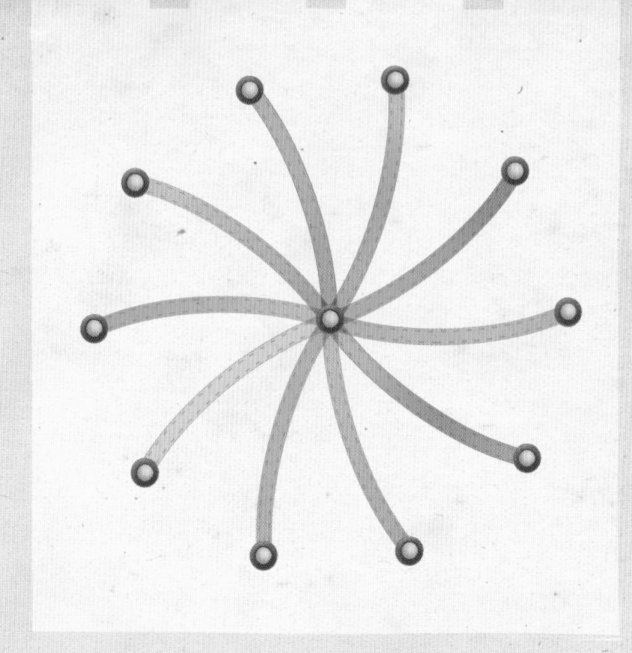

10 ribbons, 11 pearls

I was summoned to Camelot for the sake of modern art.

A festival was being prepared to honor Sir Lancelot's impressive performance in the labyrinth, and a decree was sent throughout Camelot commissioning the creation of magnificent tapestries, one for each Knight of the Round Table. These tapestries would hang around the immense pillars of Stonehenge for all to behold.

Colored ribbons of exotic silk were to be stitched onto each tapestry, with glittering pearls placed at both ends of each ribbon. The ribbons could be arranged in any pattern, but each pair of ribbons had to meet exactly once, either crisscrossing or meeting at a pearl.

o RIBBONS, 10 PEARLS

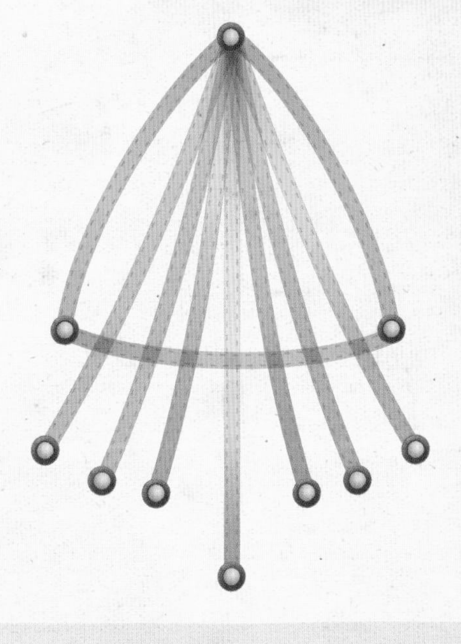

10 RIBBONS, 10 PEARLS

Although Conway the Craftsman produced numerous such beautiful tapestries to choose from, King Arthur noticed that each of them had at least as many pearls as ribbons. To honor Lancelot, Arthur asked me to design a special tapestry for him: one that conformed to the rules but would have more ribbons than pearls.

I committed myself to fashion such a tapestry before the festival began, but even with my powers of magic and logic, I was not able to succeed.

When drawing a planar depiction of a graph, some of the graph's apparent features are specific to the drawing, whereas others may indicate an intrinsic property of the graph itself. As a general rule, mathematicians want to simplify the drawing so that the connections between nodes and edges is as clear as possible. That, however, is not the goal in this story.

The tapestry designs are a special way of depicting a graph that mathematician John Conway named a thrackle. A *thrackle* is a drawing of a graph in which every pair of edges meets exactly once, either by crossing one another or at a shared endpoint. For example, the triangular graph with three nodes and three edges is a thrackle. A heptagon, a seven-sided polygon, is not a thrackle, but the graph it represents can be drawn as one.

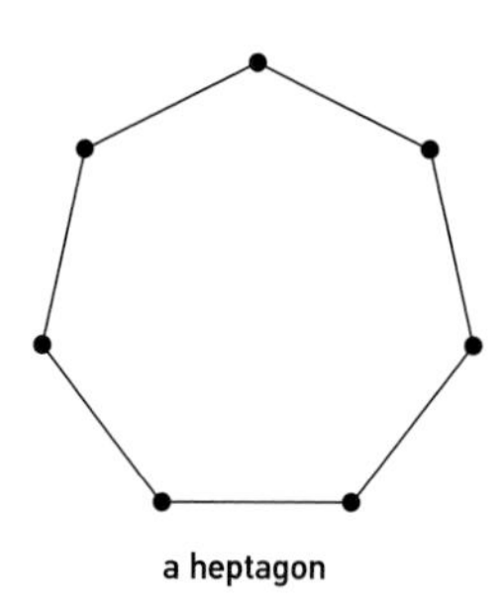

a heptagon

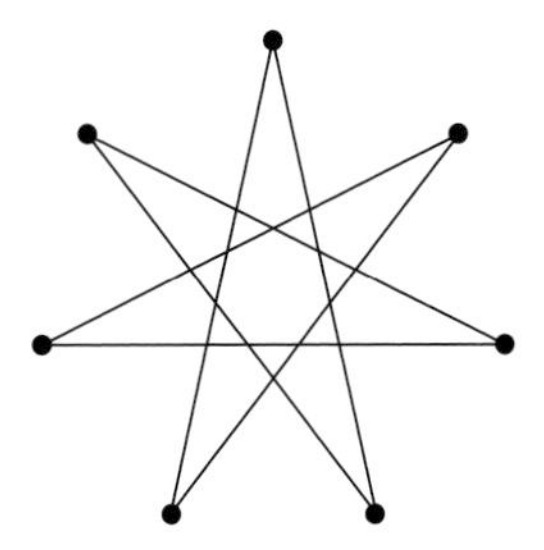

a thrackle embedding of a heptagon

It is now known that every polygon other than a quadrilateral can be represented in thrackle form, although the argument used to show this tends to yield particularly messy depictions. Merlin's question is the following conjecture of Conway:

UNSOLVED: A thrackle cannot have more edges than vertices.

Paul Erdős proved this for thrackles whose edges are all straight, but the general conjecture remains unsolved. There has been gradual improvement of the upper bound: as of 2017, it is known that a thrackle with n vertices can have no more than $1.3984n$ edges.

MYSTERY 9

mystic mountain madness

As strange and mysterious as Camelot was, could it really have been entangled with a legend as renowned as the quest for the Holy Grail? This is certainly one of the more implausible stories for my rational mind to accept. The mathematical puzzle that Merlin describes is to find a special number, one that is not the sum of two primes, just as the previous mystery was to find a special graph.

I was summoned to Camelot for a matter of utmost importance.

Rumors had reached King Arthur that the long-lost Holy Grail lay just beyond the Mystic Mountain. Unfortunately, the only path was along a narrow route beneath the mountain that branched off to create two separate exits on the other side.

Word had reached me that the warlock Goldbach was protecting the Grail, and that he had cursed with madness all those who dared to pass beneath the mountain. It was commonly known, however, that the warlock had a strange quirk, one that Arthur could use to his advantage: Goldbach was obsessed with rectangles, believing them to be shapes of absolute order and power.

Arthur therefore planned to send soldiers in a rectangular, two-column formation through this narrow route, where each column would emerge on the other side along a separate exit. Two Knights of the Round Table would accompany each column of soldiers, one in the front and one in the rear. Although these four knights would be immune to Goldbach's magic and would emerge unscathed from their designated exits, the soldiers might fall under the spell of confusion and break formation, emerging from random exits.

Arthur wondered if there was a specific number of soldiers that he could send under the mountain so he could be certain that at least one of the groups that emerged from an exit could march in a perfect rectangular formation, being found worthy of the warlock to retrieve the Holy Grail.

I was maddened by this mystery, but even with my powers of magic and logic, I was never able to answer him.

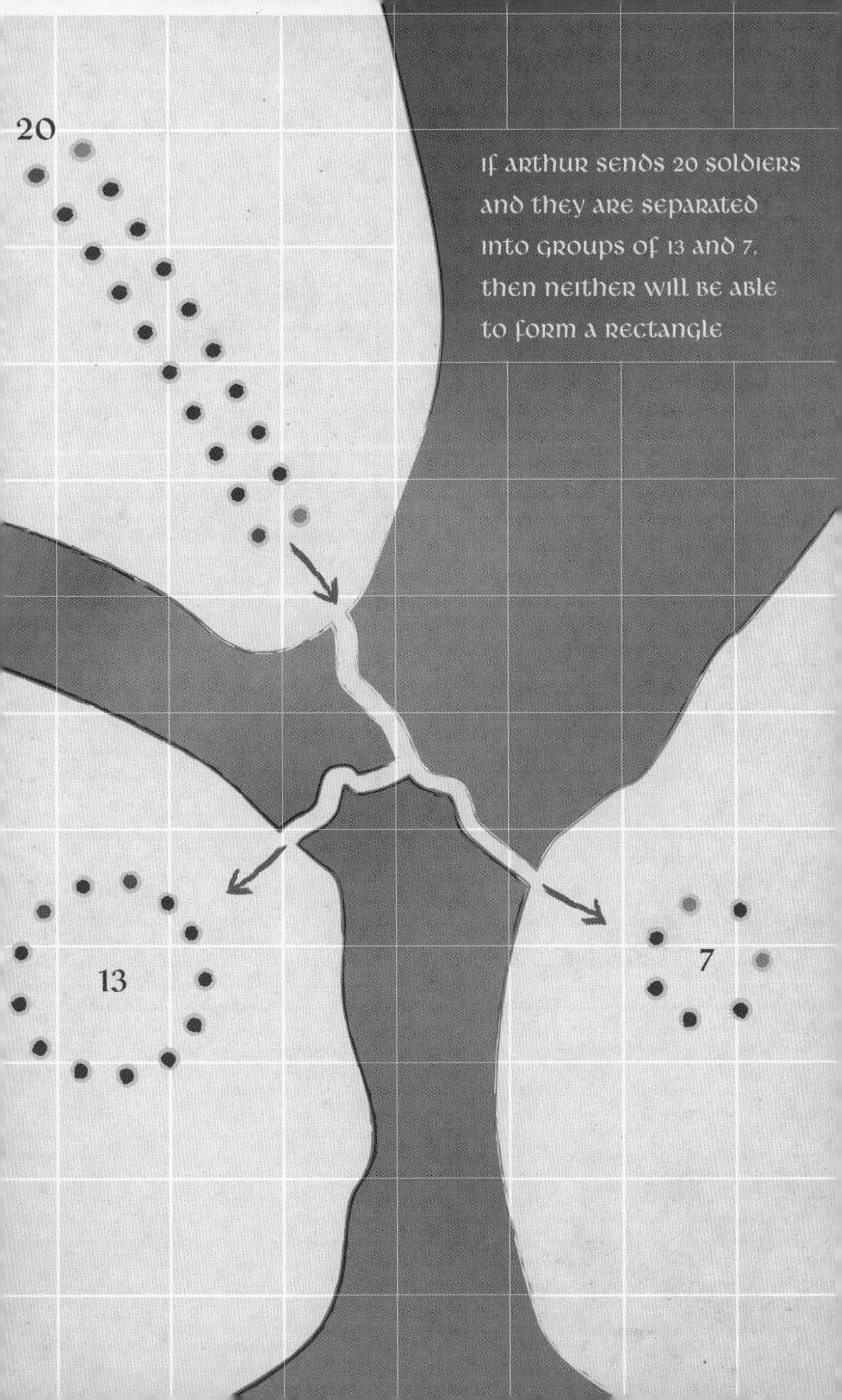
20
If Arthur sends 20 soldiers
and they are separated
into groups of 13 and 7,
then neither will be able
to form a rectangle
13
7

What considerations determine how many soldiers Merlin should send down the underground path? One requirement is that, because the soldiers are initially arranged in two columns, he must send an even number of soldiers. The other requirement is more challenging: Merlin wants to be assured that, after the soldiers are separated into two units, at least one of the units can form into a rectangle. That will be possible unless both units have a prime number of soldiers. For instance, if a unit has twelve soldiers, it can form a 3×4 rectangle, but if it has thirteen soldiers, it can only form a single file-line.

The problem is that although Merlin knows the total number of soldiers entering the mines, he cannot predict how they might become separated. For example, if Merlin decides to send twenty-four soldiers, and they are separated into units of fifteen and nine, then both units can form rectangles; if they are separated into units of twenty-one and three, then the first unit can form a rectangle but the second cannot. If the soldiers are separated into units of seventeen and seven, however, then neither unit can form into a rectangle.

To prevent this possibility, Merlin must find an even number that cannot be written as a sum of two primes, and this is not easy.

4 = 2 + 2	18 = 5 + 13
6 = 3 + 3	20 = 7 + 13
8 = 3 + 5	22 = 5 + 17
10 = 5 + 5	24 = 7 + 17
12 = 5 + 7	26 = 7 + 19
14 = 7 + 7	28 = 5 + 23
16 = 5 + 11	30 = 7 + 23

Continuing in this way, it begins to seem that perhaps every even number can be written as a sum of two primes. That computational evidence led Christian Goldbach in 1742 to conjecture the following:

UNSOLVED: Every even number (greater than two) can be written as the sum of two primes.

Thanks to computers, it was shown in 2013 that every even number less than 4,000,000,000,000,000,000 can be written as a sum of two primes. However, there is still no proof of Goldbach's conjecture.

MYSTERY 10

Camelot Carnival Celebration

Merlin makes one last foray into the world of graphs, this time turning his attention to a particular type of graph: the *tree*. These special classes of graphs show up in all sorts of areas that care about organizing information, from cartography to genetics to data storage. The challenge here involves carefully labeling the nodes of the trees. This was one of the puzzles that I was obsessed with as a kid. I would draw a small tree and then try to label it. It wasn't easy, and there was always a lot of erasing.

the holy grail had been recovered and brought back to camelot!

For this great honor, a monumental celebration was planned, replete with jousting, feasting, and dancing. Workers pitched tents of varying heights, which were connected by banners that were hung between the tent poles. When two tents were connected by a banner, they called the difference in the tent heights the *drop* of the banner. According to tradition, a typical celebration in Camelot called for fifteen tents, each of a different height, that were connected using fourteen banners, each with a different drop.

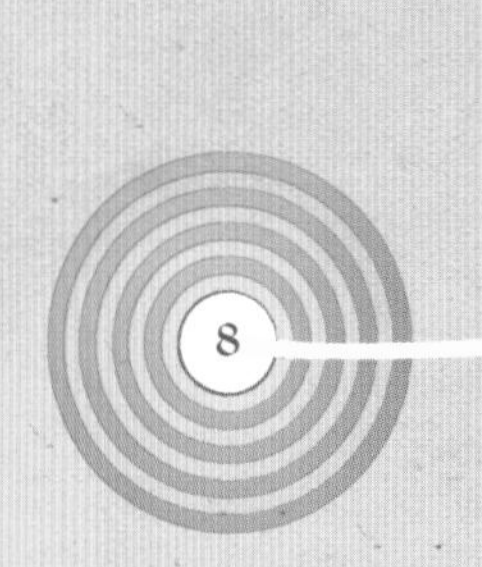

To mark the occasion of the Grail's discovery, King Arthur invited the sorceress Morgana to organize this festival in the nearby Broceliande forest. She was planning an elaborate layout of one hundred tents that would be connected together using ninety-nine banners.

Arthur asked me whether it would be possible for each tent to have a different height, ranging from 1 yard all the way to 100 yards, and for each banner to have a different drop, regardless of Morgana's final layout.

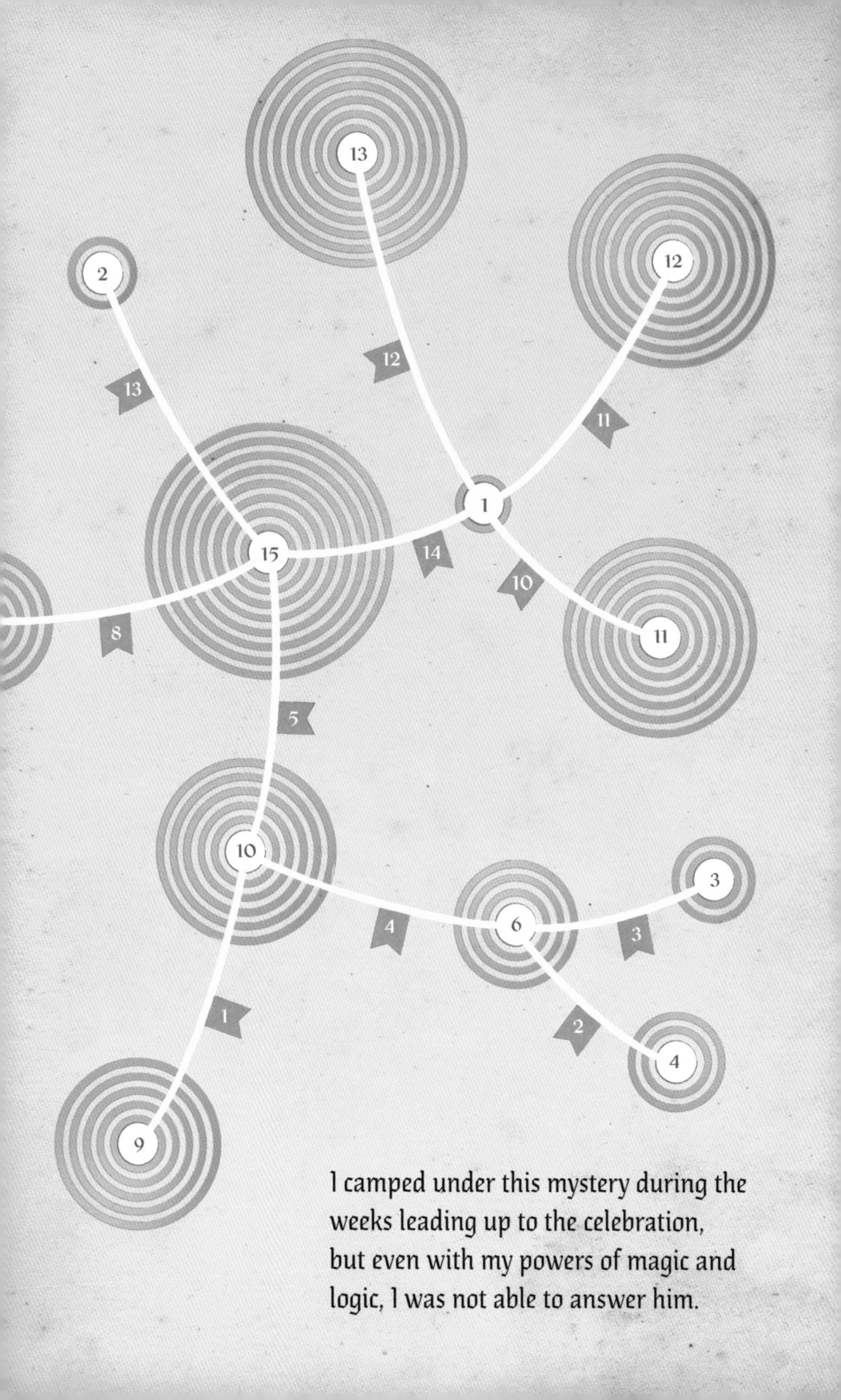

I camped under this mystery during the weeks leading up to the celebration, but even with my powers of magic and logic, I was not able to answer him.

Like a few others before it, this story describes an unsolved problem from graph theory. A graph is said to be *connected* if it is possible to travel along a chain of edges from any node to any other. A connected graph is called a tree if, for each pair of nodes, there is only one path between them. A tree always has one fewer edges than nodes. Therefore, if a tree has n nodes, its nodes can be numbered 1, 2, 3, . . . , n and its edges 1, 2, 3, . . . , $n - 1$. A tree is said to be *graceful* if it is possible to number its nodes and edges so that each edge number is equal to the difference of the node numbers that it connects.

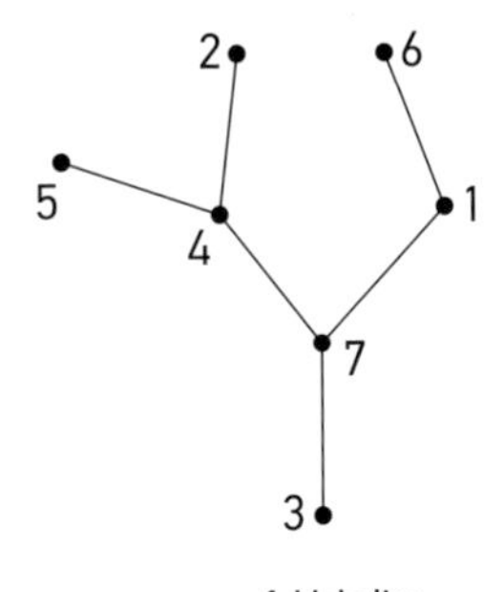

a graceful labeling

The 15-tent configuration depicted in Merlin's story shows a graceful tree: tent heights correspond to node numbers and banner drops correspond to edge numbers.

Merlin's question is whether it is possible to assign numbers to a much larger, 100-node tree so that it is graceful. (Since all the tents must be connected, and since there is one fewer banner than there are tents, Morgana's design will have to be a tree.) Numbering any particular 100-node tree would be challenging, but without advance knowledge of Morgana's design, the possibilities are nearly endless. In the early 1980s, however, Gerhard Ringel and Anton Kotzig made a prediction:

UNSOLVED: All trees are graceful.

Trees are usually amenable to "divide and conquer" type arguments: trees can be broken down into smaller trees, and results about the smaller trees can be extended to the larger ones. Unfortunately, that does not seem to be the case for this conjecture: small changes to one part of the tree tend to require large changes to the numbering throughout the tree. A distributed computing project has confirmed that all trees with fewer than 35 nodes are graceful. This is far from the 100-node trees of Merlin's carnival.

MYSTERY 11

ROWBOAT DISTRIBUTION

Mathematicians seek to abstract ideas, to observe properties in one situation and apply them to another. A simple example of this is addition: the idea of adding numbers can be extended to adding groups of numbers. Although radically different in the details, this mystery has the same flavor as the previous one: both of them work for small numbers (fifteen tents, ten boat pairs) and become improbably complicated for larger ones (one hundred tents, two hundred boat pairs).

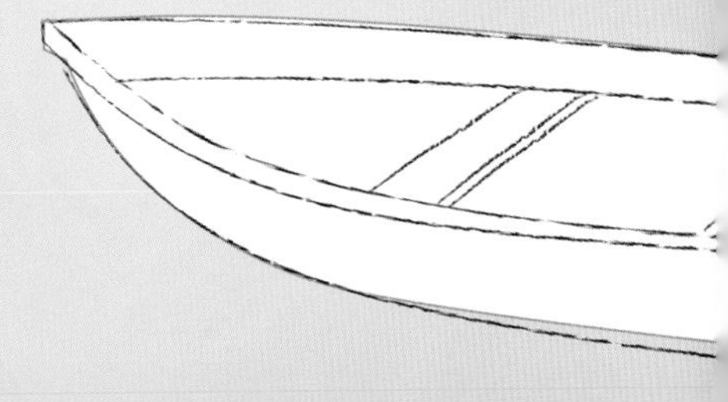

I was summoned to Camelot to help with traffic congestion.

The Lake of Camelot was a large body of water with piers of entry located at three popular sites: Stonehenge, the castle, and the mysterious isle of Avalon. There were ten rowboats and ten passenger boats of varying sizes, ranging from a one-seat rowboat with its matching one-seat passenger boat to a ten-seat rowboat with its matching ten-seat passenger boat, with every size in between.

At the end of each day, the boats were divided among the three piers for repairs and cleaning. Years ago, I devised a brilliant method to distribute these boats efficiently. First, each pier received matching rowboat and passenger boat pairs. And second, a rowboat was not allowed to have the same number of seats as any possible combination of rowboats and passenger boats put together in that pier.

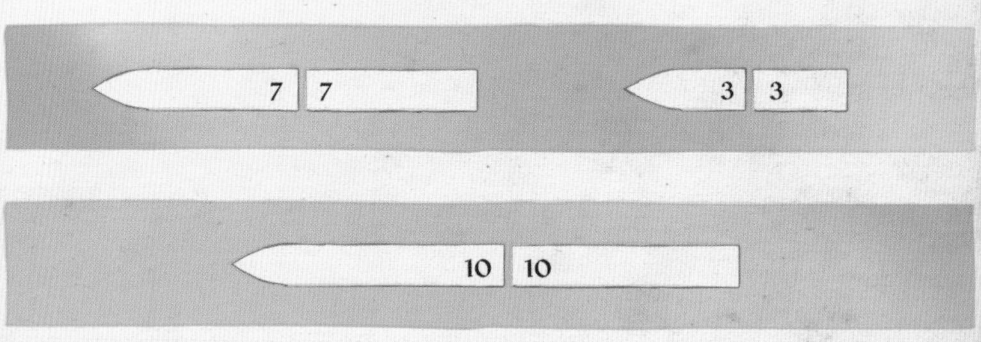

3 + 7 = 10,
so the ten-seat rowboat and passenger boat cannot dock with the three- and seven-seat boats

The diagram below shows my distribution of the boats among the three piers. Notice how the pier at Stonehenge could have combinations of boats with four, five, six, nine, ten, and fourteen seats.

Following the Grail's discovery, an influx of visitors to Camelot overburdened our boats. The king and queen decided to enlarge the system, replacing the three piers on the lake with five piers, and the ten pairs of boats with two hundred pairs. They asked me if I could distribute the two hundred boat pairs among the five piers using my special method of distribution: at each pier, a rowboat must not have the same number of seats as any possible combination of rowboat and passenger boat docked there.

I anchored myself to this puzzle, but even with my powers of magic and logic, I was not able to succeed.

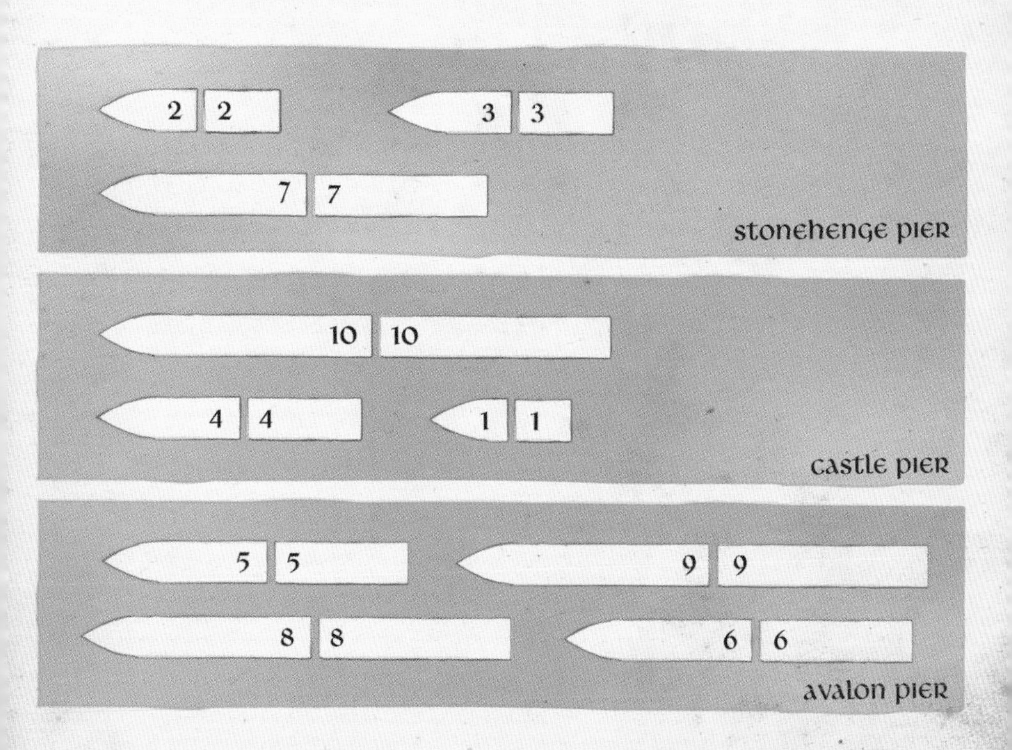

From one set of numbers S, we can form another set $S + S$ that consists of the numbers that are a sum of two numbers from set S. For example, if $S = \{1, 2, 3\}$, then $S + S = \{2, 3, 4, 5, 6\}$, since the possible sums are:

$$1 + 1 = 2$$
$$1 + 2 = 3$$
$$1 + 3 = 4$$
$$2 + 2 = 4$$
$$2 + 3 = 5$$
$$3 + 3 = 6$$

It is to be expected that S and $S + S$ may share numbers, as in this example with 2 and 3. If S and $S + S$ do not share any numbers, then S is said to be *sum-free.*

In Merlin's story, the number of seats in the rowboats are the numbers of S. The number of seats in the rowboat and passenger boat combinations are the numbers of $S + S$. For example, at Stonehenge Pier, $S = \{2, 3, 7\}$ and $S + S = \{4, 5, 6, 9, 10, 14\}$. Merlin's efficiency requirement (a rowboat and a rowboat–passenger boat combination with the same number of seats cannot dock at the same pier) guarantees that the boat numbers at each pier make up a sum-free set.

In the early twentieth century, Issai Schur studied sum-free sets. He was interested in the following question: for a particular whole number k, what is the largest possible number n so that the numbers 1, 2, 3, . . . , n can be separated into k sum-free sets? This is now called the kth Schur number, written as $S(k)$.

As an easy example, we can determine $S(2)$. Start by placing 1 into one of the two sets—call it set A. Since $1 + 1 = 2$, 2 will have to be placed in the other set—call it set B. Then 3 can be placed in either set. If 3 is placed in A, 4 cannot be placed in either A or B, since $4 = 3 + 1 = 2 + 2$. If 3 is placed in B instead, 4 can be placed in A. At this point, 5 cannot be placed in either set, since $5 = 1 + 4 = 2 + 3$, and we have run out of options. Therefore $S(2) = 4$.

As k increases, the options proliferate, and the relatively simple analysis that worked to determine $S(2)$ becomes intractable. Merlin's pier arrangement shows that $S(3)$ is at least 10, but it is actually known to be a little larger: $S(3) = 13$. It is also known that $S(4) = 44$. Beyond those few cases, exact values of $S(k)$ are unknown. A *greedy* algorithm—where a number is placed into the first set if possible, then the second, then the third, and so on—provides some lower bounds, but generally does not give the optimal answer.

UNSOLVED: Calculate the Schur numbers $S(k)$, and $S(5)$ in particular.

Merlin will be able to extend the pier and boat system if $S(5)$ is at least 200. As of 2018, all that is known is that $S(5)$ is between 160 and 315.

MYSTERY 12

holy grail vault

Physics tells us that a simple principle governs how light reflects off of smooth, flat surfaces: the angle of incidence is equal to the angle of reflection. This same principle applies to billiard balls bouncing off the edges of pool tables. This puzzle asks Merlin to think about reflections bouncing not in a rectangular room but in rooms of all possible shapes of polygons. Leave it to mathematicians to devise thorny problems from wonderfully simple principles.

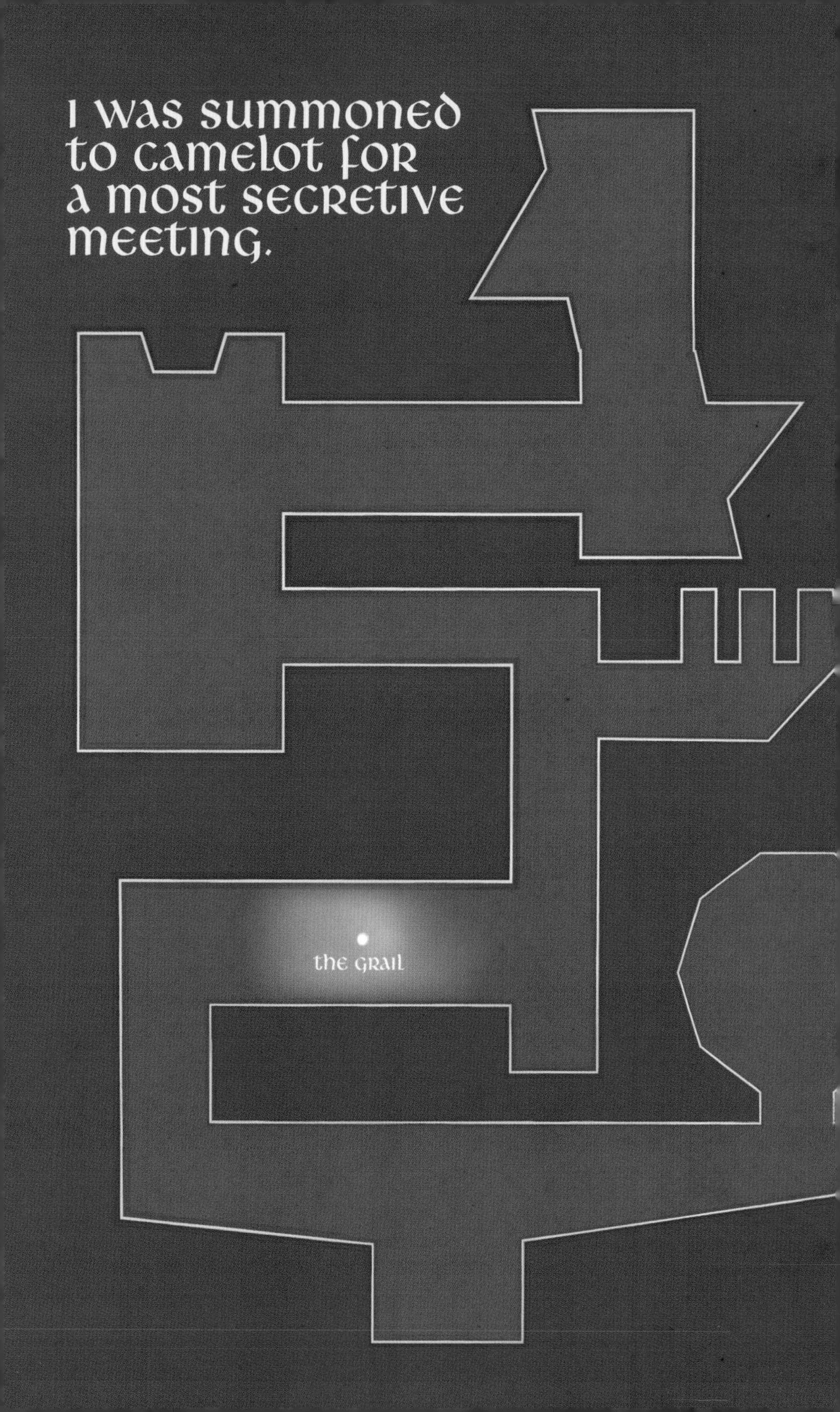
I was summoned
to Camelot for
a most secretive
meeting.
the Grail

In order to protect and safeguard the Grail, the knights wanted to construct a vast and intricate vault underneath the castle. They needed the walls of this crypt to be smooth and flat, covered with floor-to-ceiling upright mirrors. I believed this was an ingenious idea, for a mirrored room would indeed confuse and confound any would-be thief of the Grail.

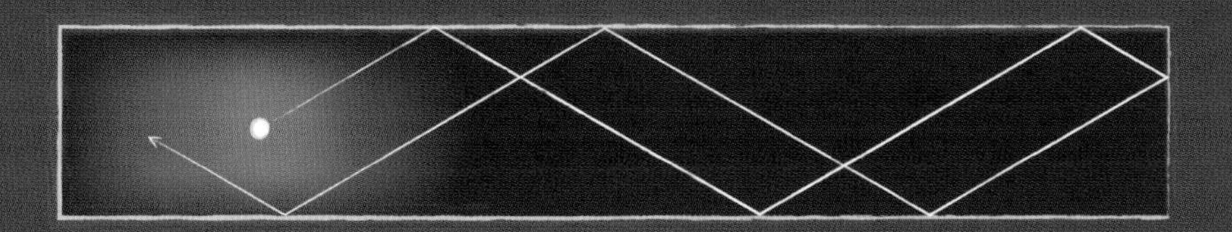

The light source to this vault would be the Grail itself, for although it looked ordinary in sunlight, the Grail emitted a powerful light in the darkness. Regardless of the final blueprint design, the knights wondered whether it would always be possible to place the Grail somewhere in the vault so that its light, bouncing off the mirrored walls, would illuminate the entire room.

I reflected on this enigma for years, but even with my powers of magic and logic, I was never able to answer them.

The Greek mathematician Hero of Alexandria first proposed the principle for light reflection, that the angle of incidence is equal to the angle of reflection.

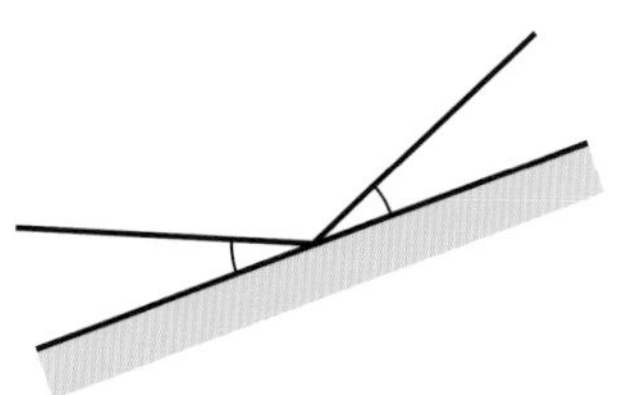

In a room with perfectly reflective walls, such as the vault in this story, a single ray of light will bounce indefinitely from wall to wall, its path determined by this principle.

In the 1950s, Ernst Straus asked whether a single light source will always completely illuminate a room made entirely of reflective walls. That is, will rays of light emanating from a single source cross through every point in the room? Even without mirrors, a single source will completely light a convex room, and just a few bounces will take light beams around simple corners. Thus, if a room cannot be completely lit, it must be built with a devious configuration of nooks and crannies to keep the light beams out.

UNSOLVED: Any room with flat mirrored walls can be fully illuminated by a single light source.

Shortly after Straus posed his problem, Roger Penrose (the same physicist mentioned in "Round Table Tiles") found a counterexample by constructing a room with elliptical walls that would prevent total illumination no matter where the light source was positioned.

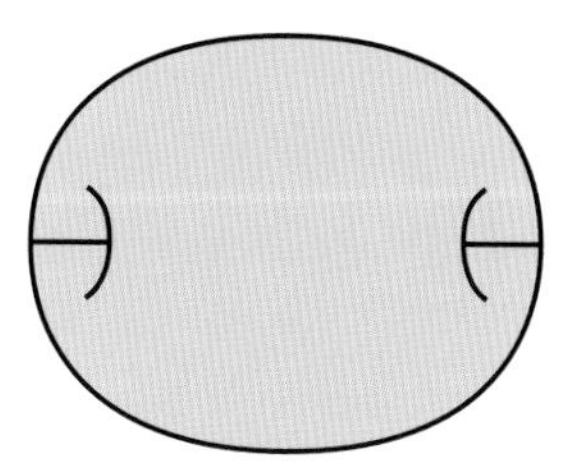

Penrose's room

In the 1990s, George Tokarsky made a bit more progress. He provided an example of a polygonal room containing two points positioned in such a way that a light placed at one point will not illuminate the other.

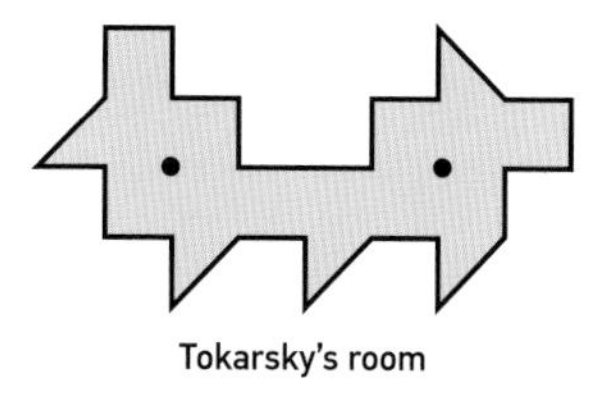

Tokarsky's room

If the light is moved away from those two points, however, it will illuminate the entire room, and so again this does not exactly meet the vault specifications.

MYSTERY 13

Starstone Shattering

In this journal entry, there is a suggestion of a mysterious jewel that is as famous as the Holy Grail. Even if it really existed, Arthur's desire to fragment it (even for the noble sake of equity) feels shortsighted in my eyes. Yet the mathematics revealed is one of beauty. Instead of dividing numbers evenly, the riddle revolves around dividing shapes.

I was summoned to Camelot as an advocate of equity.

With the successful recovery of the Holy Grail, Arthur began a new quest for another relic shrouded in mystery—the long lost Starstone. Only three things were known about this arcane jewel:

1. It was harder than a diamond but thinner than a piece of parchment.
2. It had perfectly straight sides.
3. All its corners jutted out.

Arthur promised his twelve knights that when the Starstone was recovered it would be shattered into twelve smaller jewels, each one featuring straight sides and corners that jutted out. Not knowing the exact shape of the Starstone, Arthur wondered whether this could always be done fairly, so that every one of the smaller jewels had the same area and the same perimeter.

Despite the Starstone never being found, I dissected this mystery for years, but even with my powers of magic and logic, I was never able to answer him.

valid shatterings of three possible starstone shapes

Just as numbers can be divided into equal pieces, so can shapes. A *fair partition* of a polygon is a way to cut it into pieces so that all the pieces have the same area and perimeter. It is easier to see some fair partitions than others. For instance, it is easy to cut a regular octagon into eight pieces with equal area and perimeter. But it is not so easy to cut it into nine or ten pieces. Eleven pieces would be even more of a challenge.

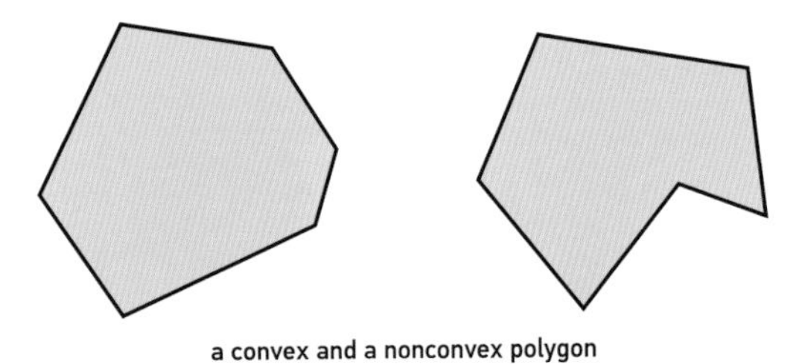

a convex and a nonconvex polygon

An arbitrary polygon becomes too difficult to handle, and so we focus on convex polygons. A polygon is convex if there exists an unbroken line of sight between any two points inside it; that is, all of its corners jut out. In 2007, R. Nandakumar and N. Ramana Rao made the following conjecture:

UNSOLVED: Any convex polygon can be fairly partitioned into n convex pieces, for any whole number n.

We know that it is possible to fairly divide any convex polygon into two pieces. The argument is as follows: Let p and q be points on the edges of the polygon so that pq cuts the polygon into two pieces (A and B) with equal areas. It is likely that the pieces do not have the same perimeter. Say, for instance, that the perimeter of piece A is less than that of piece B. Now begin to move *p* clockwise around the edge of the polygon. In order for the two pieces to retain equal area, *q* must also move clockwise. Continuing in this way, eventually point *p* will cycle around to where *q* began, and at that moment *q* will be where *p* began. Pieces A and B have switched sides as well, and so now piece A must have the greater perimeter. Throughout the process, the perimeters change continuously, so somewhere between the start and the end of the clockwise movement, the two perimeters must be equal.

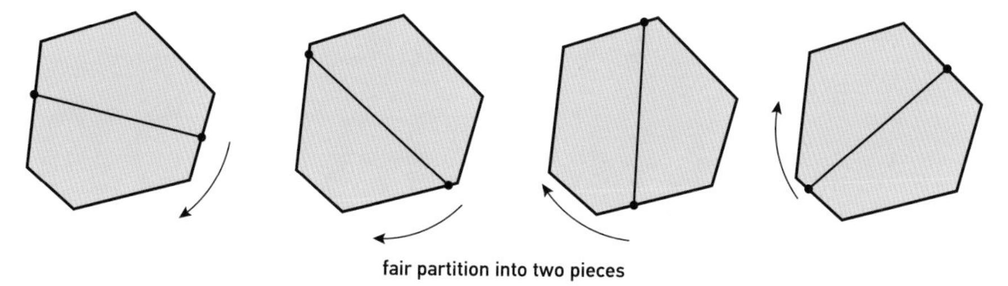

fair partition into two pieces

Although the more general problem of cutting into *n* pieces remains unsolved, some progress has been made. In 2010, Boris Aronov and Alfredo Hubard and, independently, Roman Karasev, showed that it is always possible to fair partition into *n* convex pieces if *n* is a prime number or a power of a prime number. Merlin's problem is to fairly divide a convex polygon into twelve pieces. Since twelve is not a prime number, or a power of a prime number, that result does not directly address this problem. Using Karasev's method, it is possible to fairly divide the polygon into four pieces, and then each of those into three pieces. The twelve resulting pieces will have the same area, but may not have the same perimeter.

MYSTERY 14

Knights Square Formation

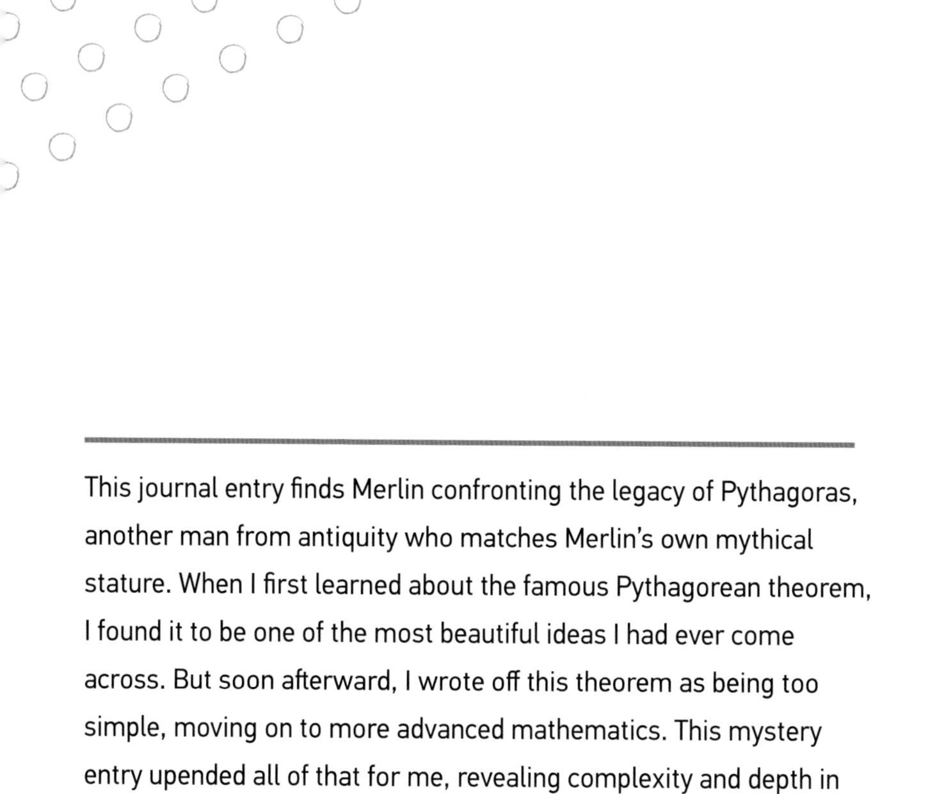

This journal entry finds Merlin confronting the legacy of Pythagoras, another man from antiquity who matches Merlin's own mythical stature. When I first learned about the famous Pythagorean theorem, I found it to be one of the most beautiful ideas I had ever come across. But soon afterward, I wrote off this theorem as being too simple, moving on to more advanced mathematics. This mystery entry upended all of that for me, revealing complexity and depth in classic ideas.

I was summoned to Camelot for the defense of the kingdom.

225 SOLDIERS

64 SOLDIERS

289 SOLDIERS

Princess Vivian had betrayed us all! Fueled by the support and dark counsel of Morgana, Vivian forged an army of her own, gathering a legion of knights in the expansive Field of Camlann, challenging Arthur for the throne of Camelot.

An infantry of soldiers in Camelot was always arranged in a square formation, signifying order and perfection. There were two infantry regiments in Camelot, one with 64 soldiers commanded by Sir Percival's and another with 225 soldiers led by Sir Galahad. Arthur deployed the regiments separately, and for serious situations, had combined them into a single unit of 289 soldiers.

With Morgana by her side, Arthur knew that Vivian's army would be formidable. In hopes of defending Camelot, Arthur reformed his forces into a new *three*-infantry army of soldiers, led by his brave knights Percival, Galahad, and Lancelot.

Arthur wanted to know how many soldiers to recruit to create an army so that each of the three infantries, any pair of the three, and all three together could march in perfect square formations.

I attacked this matter with all my might during the weeks leading up to the great battle, but even with my powers of magic and logic, I was not able to answer him.

From antiquity, mathematicians of many cultures have been aware of the relationship between the lengths of the three sides of a right triangle, which we now call the Pythagorean theorem.

PYTHAGOREAN THEOREM: If a and b are the lengths of the legs of a right triangle, and c is the length of its hypotenuse, then $a^2 + b^2 = c^2$.

Because of this relationship, it is unusual for the lengths of all three sides to be whole numbers. For instance, if both legs have a length of 1, then the hypotenuse has a length of $\sqrt{2} = 1.414 \ldots$. There are, however, some right triangles whose side lengths are all whole numbers; for example, there is a right triangle with sides measuring 3, 4, and 5 in length, since

$$3^2 + 4^2 = 9 + 16 = 25 = 5^2.$$

Such a trio of side lengths is called a Pythagorean triple. The most direct way to search for a Pythagorean triple is to pick two whole numbers, a and b, and to check whether $a^2 + b^2$ is a perfect square. In Merlin's story, the existing troop arrangement provides an example: with $a = 8$ and $b = 15$,

$$a^2 + b^2 = 64 + 225 = 289 = 17^2.$$

Merlin's challenge is to pick three whole numbers—a, b, and c—and to check whether four combinations of them are perfect squares: $a^2 + b^2$, $a^2 + c^2$, $b^2 + c^2$, and $a^2 + b^2 + c^2$. There is a geometric way to understand this: if a, b, and c are the lengths of the sides of a rectangular box, then $\sqrt{a^2 + b^2}$, $\sqrt{a^2 + c^2}$, and $\sqrt{b^2 + c^2}$ are the lengths of the diagonals of its sides, and $\sqrt{a^2 + b^2 + c^2}$ is the length of the diagonal from a corner of the box to the opposite corner. If Merlin could find numbers that meet the four conditions, then all of those diagonals would be whole numbers, creating what is called a perfect brick.

UNSOLVED: Find the dimensions of a perfect brick.

This is sometimes called a perfect Euler brick to honor 18th-century Swiss mathematician Leonhard Euler, who worked on this problem. It is not too hard to find numbers that satisfy the first three requirements. For instance, if $a = 44$, $b = 117$, and $c = 240$, then

$$a^2 + b^2 = 15{,}625 = 125^2,$$
$$a^2 + c^2 = 59{,}536 = 244^2,$$
$$b^2 + c^2 = 71{,}289 = 267^2.$$

However, since $a^2 + b^2 + c^2 = 73{,}225$ (which is not a perfect square), the corner-to-corner length of this box is not a whole number.

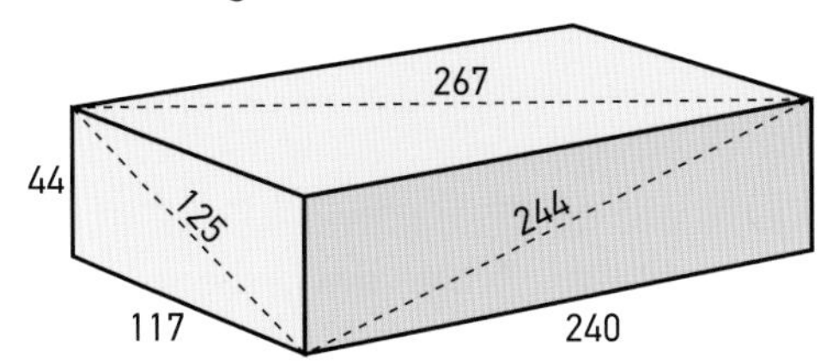

Working from the other direction, it is not too hard to write a number as a sum of three squares. Joseph-Louis Lagrange, Euler's successor as the director of mathematics at the prestigious Prussian Academy of Sciences, proved that every positive integer can be written as a sum of four or fewer squares. The challenge is to find values of a, b, and c that will satisfy all four conditions simultaneously. None have yet been found. That being said, no argument has yet been presented for why there can be no such combination of numbers.

MYSTERY 15

thirty-three tree tomb

With the tragic death of his friend Arthur, Merlin finds himself with another puzzle. It doesn't deal with primality of numbers or structures of graphs, but something far simpler: placing dots on a piece of paper. Mathematicians situate this type of question in the world of arrangements: Merlin's very first mystery deals with arrangements of six squares whereas this riddle wrestles with arrangements of thirty-three dots. I spent many nights doodling dots in search of a solution.

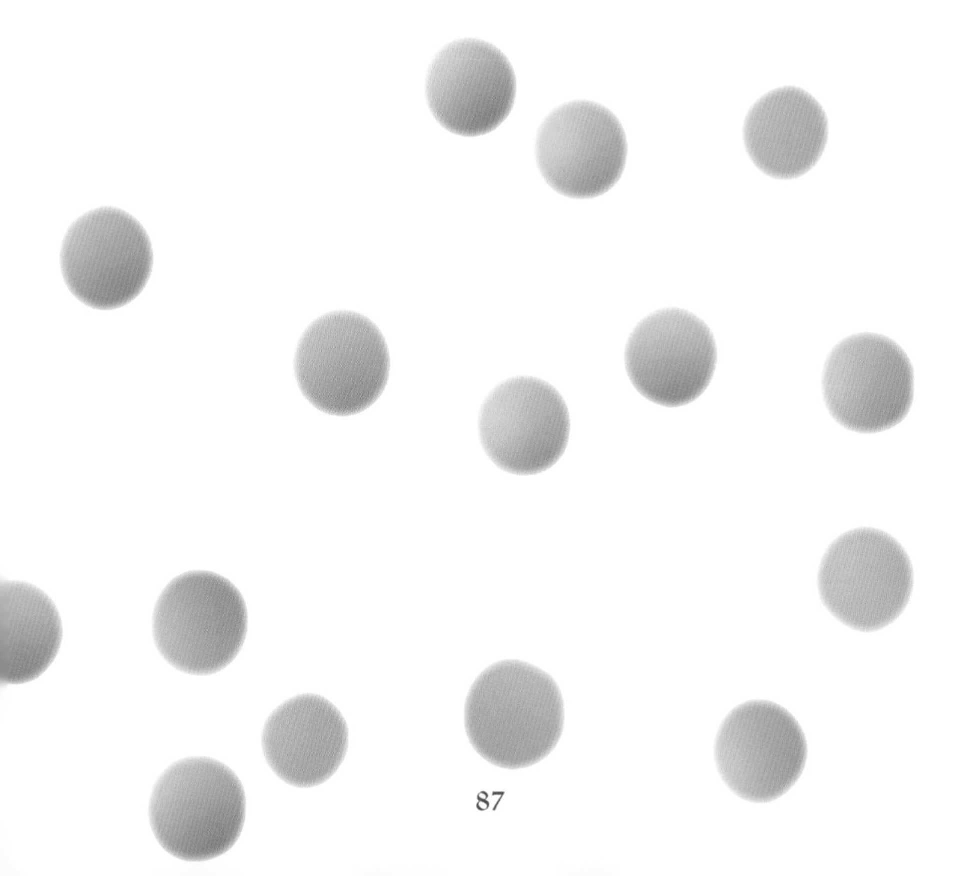

I was summoned to Camelot in a time of heartbreak and honor.

King Arthur was mortally wounded during the Battle of Camlann. He called me into his presence one last time, for his dying wish was to be buried at his birthplace, the Isle of Avalon.

Arthur told me that on this island, there was a large raised mound that was both flat and wide. And on this mound, 33 mighty oak trees grew, no three in a row.

He instructed me to choose seven of the trees, and build towers upon them to mark the corners of his glorious tomb, making sure the tomb towers will all jut out.

Not having seen the mound nor the exact configuration of trees on it, I thought that surely there must be seven that would satisfy his request. But even with my powers of magic and logic, I could find no reason why this must be so.

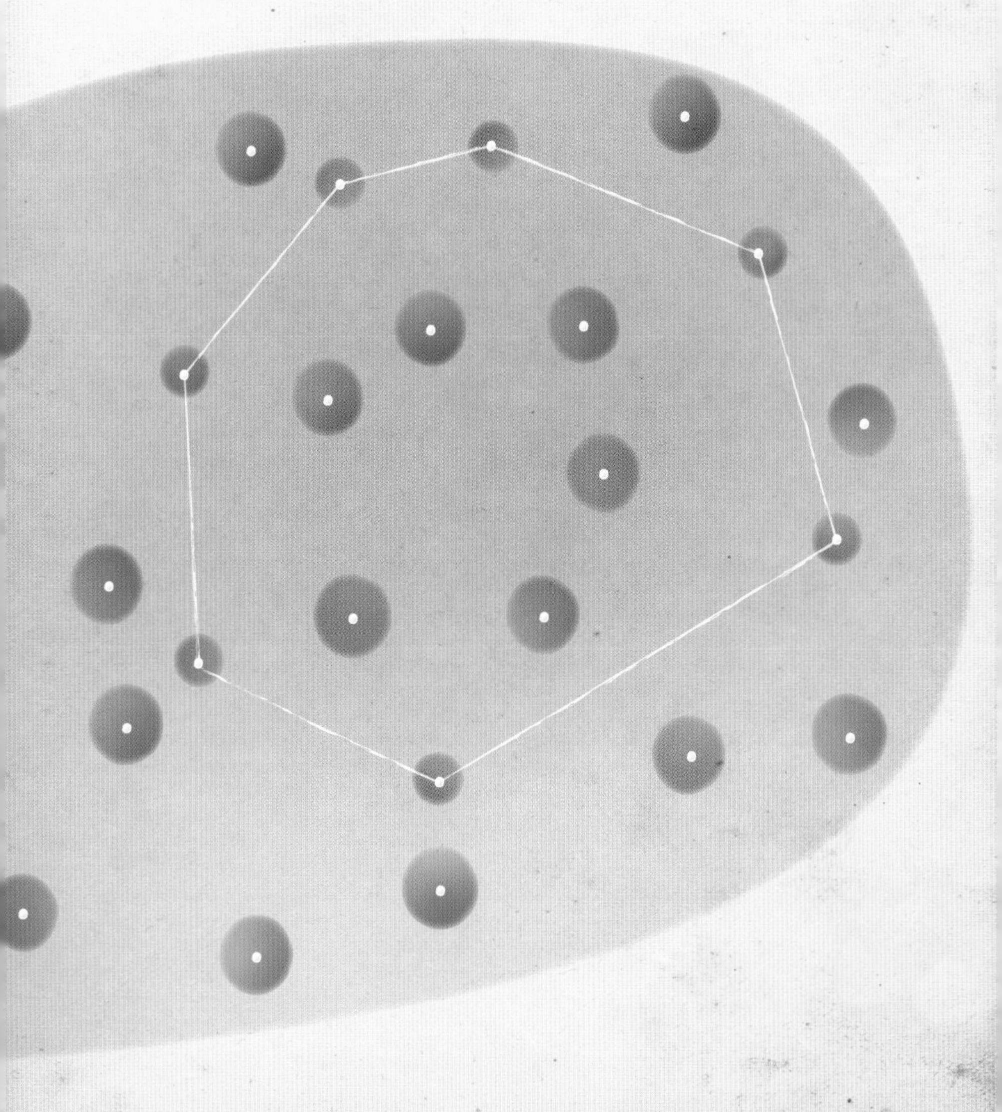

In the 1930s, Budapest was a thriving environment for young mathematicians. It was in this environment that Esther Klein presented a problem to her colleagues: given five points in general position on the plane, can you guarantee that four of them will be the corners of a convex quadrilateral? The technical requirement that the points be in "general position" simply means that no three of them are exactly in a line. Certainly, any four points in general position will serve as vertices of some quadrilateral, but it may not be convex.

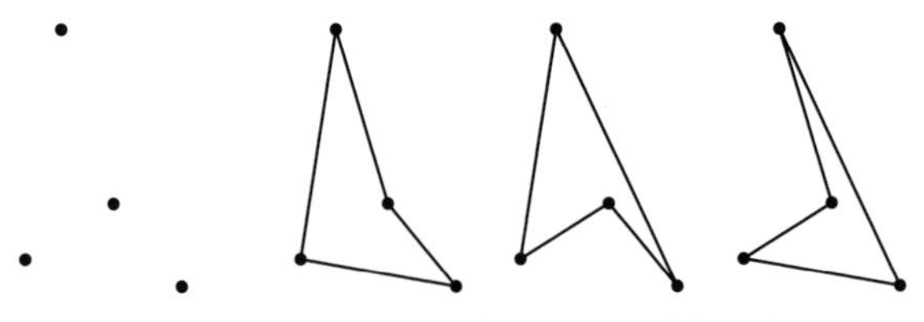

four points that are not corners of a convex quadrilateral

Klein had a solution—yes, five points do suffice—but as with many a good solution, it opened more doors. If five points are needed to guarantee a convex four-sided polygon, how many are needed to guarantee a convex n-sided polygon?

Decades later, Paul Erdős and George Szekeres proved that at least $2^{n-2}+1$ points are needed to guarantee a convex polygon, and conjectured the following:

UNSOLVED: At most $2^{n-2}+1$ points in general position are needed to guarantee a convex polygon with n sides.

Their conjecture helped open a new field of mathematics, now called combinatorial geometry. Erdős later called this the "happy ending" problem, not in relation to any of its mathematical aspects but rather because Klein and Szekeres were married a few years later.

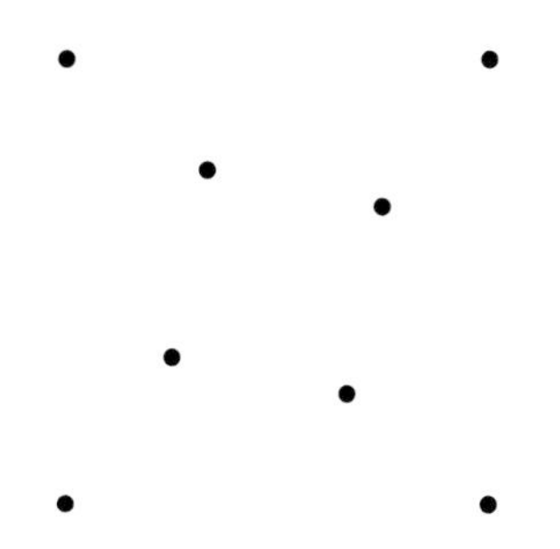

eight points that are not corners of a convex pentagon

Little progress on the problem has been made since. It has been shown that $2^3 + 1 = 9$ points are needed to guarantee a convex pentagon, and that $2^4 + 1 = 17$ points are needed to guarantee a convex hexagon. However, it is not known how many points are needed to guarantee a convex polygon with more than six sides. Merlin's challenge is to find the corners of a convex seven-sided polygon in a collection of thirty-three points. If the conjecture of Erdős and Szekeres is true, then $2^5 + 1 = 33$ trees is sufficient.

MYSTERY 16

Lady of the Lake

This is the last entry in Merlin's journal. This final puzzle is an example of *iteration*, the continual repetition of a process in which the answer to each step is then the starting point of the next one. This technique frequently appears in mathematics, where even the simplest processes yield unexpected complexity. Although this mathematical riddle still remains today, Merlin seems to realize that the Camelot he knows is coming to an end.

I was summoned to Camelot for the very last time.

The Lady of the Lake, the powerful enchantress who forged the mighty Excalibur, sought my presence. With the death of King Arthur, she foretold that the downfall of Camelot had begun. Yet in this moment of despair, she offered a path to preserve the memory of Camelot for all of time.

All she asked of me was to choose a whole number.

At the stroke of midnight, the lady would take this number and alter it to get a new number for tomorrow: If my number was even, she would replace it with half its value. And if it was odd, she would replace it with one more than triple its value.

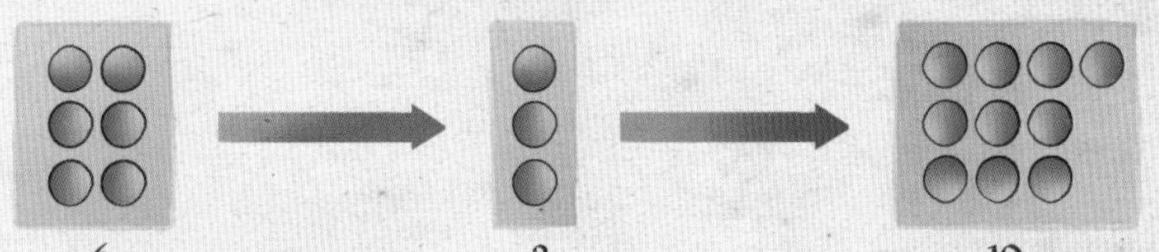

The next night, at midnight, she would alter tomorrow's number using her same rule, producing a new number for the day after. She promised me that this numbering of days would forever continue, all based on my initial choice, as long as a

9 ◆ 28 ◆ 14 ◆ 7 ◆ 22 ◆ 11 ◆ 34 ◆ 17 ◆ 52 ◆

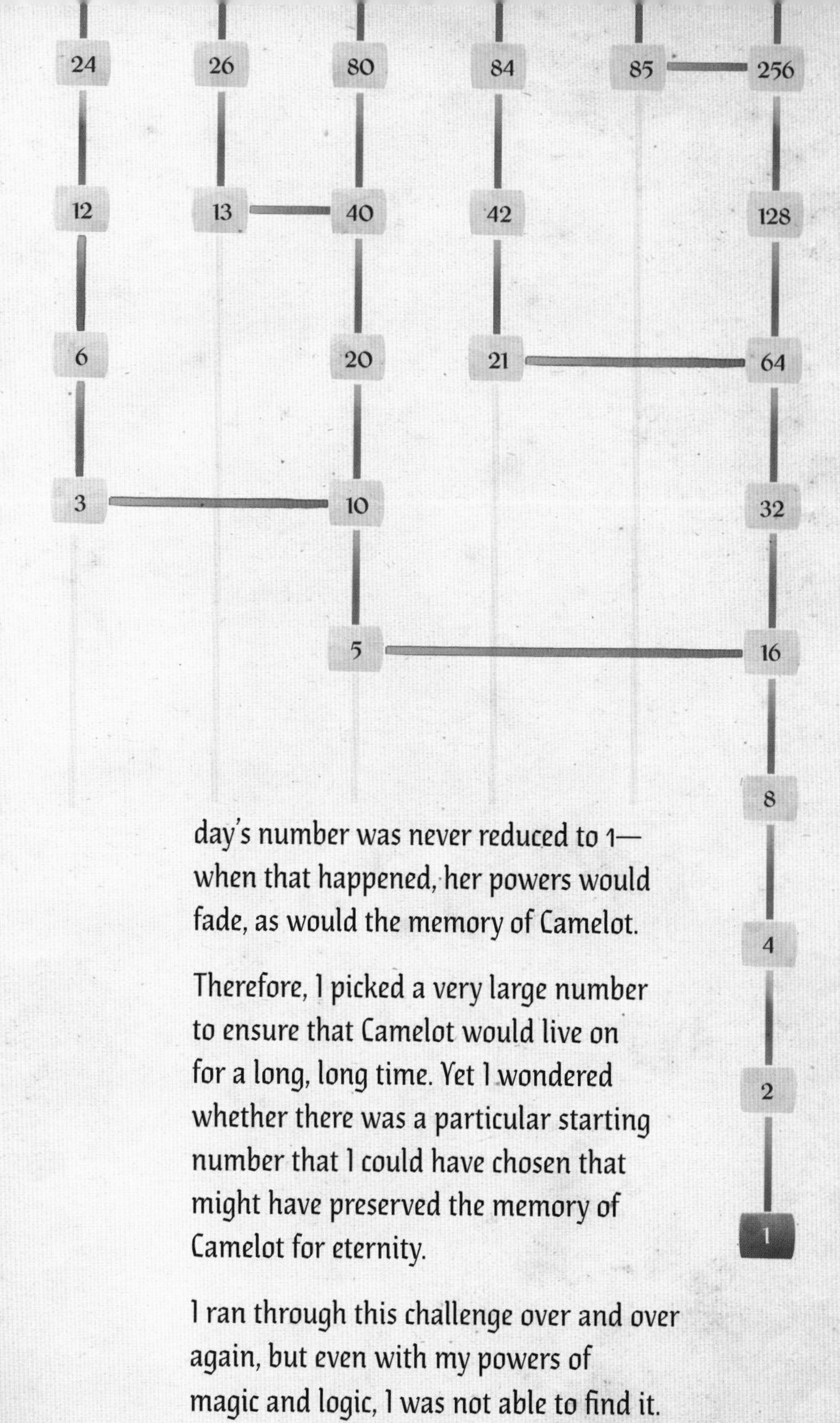

day's number was never reduced to 1—when that happened, her powers would fade, as would the memory of Camelot.

Therefore, I picked a very large number to ensure that Camelot would live on for a long, long time. Yet I wondered whether there was a particular starting number that I could have chosen that might have preserved the memory of Camelot for eternity.

I ran through this challenge over and over again, but even with my powers of magic and logic, I was not able to find it.

◆ 40 ◆ 20 ◆ 10 ◆ 5 ◆ 16 ◆ 8 ◆ 4 ◆ 2 ◆ 1

In 1937, Lothar Collatz proposed an elegant set of rules to iterate: Start with a whole number. If it is even, divide it by two; if it is odd, multiply by three and add one. The process requires only the basic arithmetic of addition, multiplication, and division, and only directly references the three smallest whole numbers: 1, 2, 3.

And yet, iteration of this algorithm yields unpredictable patterns, bouncing up to large numbers, then falling back down, back up again, and then back down. For instance,

$12 \to 6 \to 3 \to 10 \to 5 \to 16 \to 8 \to 4 \to 2 \to 1 \to 4 \to 2 \to 1 \to 4 \to$

In spite of the complexity, Collatz conjectured that every number will, upon iteration, eventually fall down to the $4 \to 2 \to 1$ cycle. Since 1 is odd, the next step is $3(1) + 1 = 4$, and from that point, further iterations cycle through the pattern $4 \to 2 \to 1$.

UNSOLVED: Upon iteration of the Collatz algorithm, every whole number will eventually arrive at 1.

There are two ways this conjecture can be false. One is if the starting number produces a sequence that increases without bound. The other is if some sequence eventually enters a cycle other than $4 \rightarrow 2 \rightarrow 1$. Jeffrey Lagarias has shown that if there is such a cycle, it must have over 275,000 terms. As of today, no such sequences have been found.

To keep the memory of Camelot alive forever, Merlin would have to find a number that does not fall into the $4 \rightarrow 2 \rightarrow 1$ pattern. Thanks to computer verification in 2017, it is now known that the conjecture is true for all starting numbers less than 87×2^{60}. In spite of this enormous evidence, a proof seems to fall well outside the boundaries of contemporary mathematics. As Paul Erdős, who figures prominently in the notes on a few of our stories, famously stated, "Mathematics is not yet ready for such problems."

ACKNOWLEDGMENTS

From the moment of this book's conception, we knew that we wanted to create something unusual, something that colored outside the lines of traditional mathematics. Many friends and colleagues have influenced and shaped our vision and understanding of mathematics over the years. For this book, we are especially grateful to Colin Adams, Dwight Bean, Art Benjamin, Ed Burger, Andy Crouch, Mako Fujimura, Tom Garrity, John McCleary, Matt Nix, Joe O'Rourke, Lior Pachter, Mike Shulman, and Lynell Weeg.

This book would not have been possible without the relentless work of our protector, Karen Gantz. We are also indebted to Jermey Matthews, Molly Seamans, and the team at the MIT Press, who believed in the project from the very beginning and helped mold our ideas into a book that matched our imagination.

Finally, gratitude and love to our families, who have supported us on this journey, as on so many others, as we make our way through life's mysteries.

FURTHER READING

Adams, Colin. *The Knot Book: An Elementary Introduction to the Mathematical Theory of Knots*. Providence, RI: American Mathematical Society, 2004.

A lovely and accessible textbook that delves into the mathematics of shape as seen through the lens of knots, graphs, and surfaces. It lifts the hood to show motivation, theory, proofs, and applications for a general audience.

Aigner, Martin, Günter Ziegler, and Karl H. Hofmann (illustrator). *Proofs from THE BOOK*, 6th ed. Berlin: Springer, 2018.

The mathematician Paul Erdős imagined that God holds a book containing the very best proof of every theorem. This unique and award-winning book, now in its sixth edition, illustrates what mathematicians mean when they say a proof is beautiful.

Benjamin, Arthur, Gary Chartrand, and Ping Zhang. *The Fascinating World of Graph Theory*. Princeton, NJ: Princeton University Press, 2015.

There are numerous books on graph theory at all levels. This is an engaging introduction to the fundamental concepts of this field that explores a plethora of classic problems and numerous exercises to engage the reader.

Caldwell, Chris. *The Prime Pages*. https://primes.utm.edu.

This website is an up-to-date repository of information about the search for large prime numbers, and other curiosities related to primes.

Conway, John H., and Richard K. Guy. *The Book of Numbers*. New York: Copernicus, 1996.

A joyful and inviting book about the origins, patterns, and interrelationships of different numbers. The clever diagrams and pictures make it even more appealing.

Doxiadis, Apostolos. *Uncle Petros and Goldbach's Conjecture*. London: Bloomsbury Press, 2000.

A charming and brilliantly written novel surrounding the pursuit of unsolved mathematics, written as a detective story and intellectual thriller that offers glimpses of remarkable beauty.

Euclid. *Euclid's Elements*. Ed. Dana Densmore, trans. T.L. Heath. Santa Fe, NM: Green Lion Press, 2007.

The grandfather of all mathematics books, a compendium of the mathematical knowledge of the ancient Greeks. Two thousand years later, it still informs and inspires contemporary mathematicians.

Grünbaum, Branko, and G. C. Shepherd, *Tiling and Patterns*, 2nd ed. Dover Books, 2017.

The encyclopedic standard book on tiling and patterns. The mathematics is quite deep, but the pages upon pages of more than five hundred illustrations are welcoming to all.

Lagarias, Jeffrey C., ed. *The Ultimate Challenge: The 3x+1 Problem*. Providence, RI: American Mathematical Society, 2011.

This book is a collection of research articles from the last forty years compiled by the leading expert on the Collatz Conjecture. It approaches the problem from different fields and different strategies, all with the intent of trying to solve this mystery.

Lakatos, Imre. *Proofs and Refutations: The Logic of Mathematical Discovery*. Cambridge: Cambridge University Press, 1971.

A unique book that is written as a dialogue between fictional students and their teacher as they try to discover and prove a famous formula of Leonhard Euler. It wonderfully shows that the creation of mathematics is an evolving, dynamic process.

Mumford, David, Caroline Series, and David Wright. *Indra's Pearls: The Vision of Felix Klein*. Cambridge: Cambridge University Press, 2006.

A visually attractive book that illustrates the complexity that can occur when simple processes are iterated.

O'Rourke, Joseph. *How to Fold It: The Mathematics of Linkages, Origami, and Polyhedra*. Cambridge: Cambridge University Press, 2011.

This accessible book is full of pictures and provides a fantastic introduction to the mathematics of origami folding, showcasing its remarkable theorems and opening the door to the launch of a new field of mathematics.

Singh, Simon. *Fermat's Enigma: The Epic Quest to Solve the World's Greatest Mathematical Problem*. New York: Anchor Books, 1998.

Based on an award-winning documentary film, this is a gripping account of Andrew Wiles's quest for the solution of Fermat's Last Theorem, filled with tales of heartbreak and wonder.

Weeks, Jeffrey. *The Shape of Space*, 3rd ed. Boca Raton, FL: CRC Press, 2020.

One of the most intuitive introductions to the ideas behind the mathematics of 2-D and 3-D objects, gently taking the reader from the simplest of pictures to sophisticated mathematics found in advanced college courses.

Wenninger, Magnus. *Polyhedron Models*. Cambridge: Cambridge University Press, 1971.

A hands-on book that explains how to assemble paper models of a vast array of polyhedra—from simple Platonic solids to extraordinary complicated stellations, complete with templates for the pieces.